56 Structure and Bonding

New Developments

With Contributions by
B. Dobiáš I. R. Epstein K. Kustin
S. K. Patil V. V. Ramakrishna

With 44 Figures and 37 Tables

Springer-Verlag
Berlin Heidelberg New York Tokyo 1984

ISBN 3-540-13106-X
Springer-Verlag Berlin Heidelberg New York Tokyo
ISBN 0-387-13106-X
Springer Verlag New York Heidelberg Berlin Tokyo

Library of Congress Catalog Card Number 67-11280

Typesetting and printing: Schwetzinger Verlagsdruckerei. Bookbinding: J. Schäffer, Grünstadt.
2152/3140-543210

Table of Contents

Design of Inorganic Chemical Oscillators

Irving R. Epstein and Kenneth Kustin

Department of Chemistry, Brandeis University, Waltham, MA 02254, USA

In a homogeneous chemical reaction maintained far from equilibrium where the reaction mechanism contains coupled feedback steps, oscillations in the concentrations of certain intermediates are possible. Until recently, the few known chemical oscillators were discovered accidentally, were variants of these reactions, or were extracted from biological systems. To provide new chemical oscillators for theoretical and experimental study, a program has been developed to design homogeneous oscillators based on reactions of inorganic compounds and ions. Mechanistic detail is presented for the chlorite-iodide-arsenite oscillator, and other newly designed oscillators are discussed more briefly. A classification scheme for grouping these and related oscillators is presented.

I. Introduction

The phrase "laboratory curiosity" was an apt characterization of a reaction that first saw the light of day in the late 1950's[1]. This reaction – the acidic oxidation of citric acid by bromate in the presence of the dual catalysts bromide and cerium(IV)/(III) – displays oscillations in the concentrations of two component species in the course of proceeding towards completion. Curiosity and skepticism were engendered by oscillation in a homogeneous reaction mixture, even though such observations had been well documented in the past.

For example, early in this century, Morgan observed pulses of carbon monoxide emanating from the decomposition of aqueous acidic formic acid[2]. Some years later, Bray began reporting studies of another decomposition reaction, that of hydrogen peroxide in the presence of iodate and iodine, observing periodicities in the concentration of iodine and in the rate of oxygen evolution[3]. These phenomena show temporal oscillation. Spatial periodicities were reported by Liesegang in 1905 to occur during the formation of inorganic precipitates such as Ag_2CrO_4[4].

Temporal and spatial periodicities occur frequently in nature. Biorhythms and biological clocks are readily detectable, periodic variations in chemical components of biological systems; e.g., firefly signaling, brain waves, etc.[5] Repetitive spatial structures like Liesegang rings are exhibited in geological layering[6], in multiple ring structures around planets such as Saturn[7], and in slime mold aggregation[8]. To this list may also be added both temporal and spatial periodicities occurring in gaseous flames[9]. Despite their ubiquity, however, these phenomena were not intensively studied. They were suspected – especially the laboratory curiosities – of being artifactual. The attitude of chemists at the time is perhaps best described by Field and Noyes: *„A chemical system resembles a pendulum subject to such severe viscous damping that the momentum is always negligibly small;... and overshoot of the final equilibrium position is impossible"*[10].

It was this strong emphasis on the non-oscillatory nature of equilibrium that led chemists to suspect the validity of oscillating reactions. The occurrence of periodic behavior was ascribed to heterogeneities – dust particles, the presence of bubbles (the Morgan and Bray reactions involved gaseous products), and so on. However, the development of a better understanding of chemical dynamics, and new experimental findings fostered a more charitable view of oscillating reactions as a challenging field of chemistry where theory and experiment intertwine in unexpected ways.

The extension of thermodynamic theory to conditions far from equilibrium by Prigogine and his school showed that spatial and temporal organization could be sustained by evolving systems[11]. The success of this, and other theories on „instability"[12], helped to remove some of the stigma associated with chemical oscillation. Experimentally, there were two significant findings motivating a serious study of oscillations. A variant of the bromate-citric acid reaction wherein malonic acid is used as reducing substrate, and known as the BZ reaction after the discoverer, Belousov[1], and first important investigator, Zhabotinsky[13], afforded an accessible, reproducible system to demonstrate and study chemical oscillations. Second, the discovery by Chance et al. of true oscillations in NADH fluorescence from cell free yeast extracts showed that glycolysis is an oscillatory process in a homogeneous medium[14].

Glycolysis, the BZ reaction, and other such systems are very complicated reactions with several components and multi-step mechanisms. However, experimental studies carried out over the past three decades by rapid mixing, chemical relaxation, line-broadening and pulse radiolytic techniques have yielded rate constants for virtually every type of elementary reaction: acid-base, redox, metal complex, hydrolysis, addition, etc. With the advent of computational techniques which permit, among other tasks, the integration of sets of coupled differential equations[15,16], the existence of these rate constant data should enable one to propose a testable mechanism for oscillation. At least one such model exists; it is the mechanism first proposed by Field, Körös and Noyes (FKN) to explain the rich array of dynamical behavior in the BZ reaction[17,18].

Despite the widespread occurrence of oscillatory chemical behavior, the acceptance of the phenomenon, and the growing interest in the subject, as late as 1980 only two non-biological systems were available for study. These were the BZ reaction and the iodine-induced decomposition of hydrogen peroxide (renamed the Bray-Liebhafsky, or BL reaction)[3,19]. In addition, variants or hybrids of these two such as the Briggs-Rauscher (BR) reaction[20] and the uncatalyzed BZ system[21] were also available. Consequently, we felt that the design of new chemical oscillators, readily available for experimental and theoretical study, was a goal worth pursuing. The result of our response to this challenge is the discovery of numerous oscillators based on inorganc compounds and ions. In the following we describe our efforts and results by first reviewing the salient features of previously described oscillators. We consider the conditions necessary for oscillation, indicate how to design an oscillator, and describe the families of new oscillators that have been discovered in our laboratories.

II. "Older" Chemical Oscillators

Prior to our systematically designed oscillators, the "inorganic" oscillator most intensively studied was the BZ reaction. Examination of this reaction and the FKN mechanism reveals several features helpful to achieving an appreciation for the construction of a chemical oscillator. The driving force for the BZ reaction is the reduction of bromate and the associated oxidation of malonic acid (MA) to carbon dioxide. The stoichiometry is not that simple, however, since bromomalonic acid is also produced in the reaction, and can be oxidized to formic acid. A possible stoichiometry is

$$3\,BrO_3^- + 5\,CH_2(COOH)_2 + 3\,H^+ \xrightarrow{\text{catalyst}} 3\,BrCH(COOH)_2 + 2\,HCOOH + 4\,CO_2 + 5\,H_2O \quad (I)$$

though during the oscillatory part of the reaction it is more accurately represented as

$$2\,BrO_3^- + 3\,CH_2(COOH)_2 + 2\,H^+ \xrightarrow{\text{catalyst}} 2\,BrCH(COOH)_2 + 3\,CO_2 + 4\,H_2O$$

where catalyst = Ce^{3+}/Ce^{4+}, Mn^{2+}/Mn^{3+}, or other redox couple with a similar potential, and Br^-.

Although bromate is a powerful oxidizing agent, it is kinetically inert. Reaction (I) occurs at a resonable rate because of the generation of reactive bromine-containing species. At the core of the FKN mechanism lies the postulate that there are *two* distinct processes producing lower-valent bromine species, and that oscillations arise when the mechanism switches from one to the other. Process (A) is essentially attack of bromide on bromate which produces bromine, an effective brominating and oxidizing agent of organic molecules. Process (B) is the autocatalytic buildup of bromous acid due to the intervention of the redox catalyst. The sequence of reactions is shown in Table 1. When the concentration of bromide is low, the system switches from Process A to Process B. Bromide ion is produced as a consequence of Process B and (R 10), and the system switches back to Process A when the bromide level rises to the point where it competes favorably for bromous acid. The presence of at least one autocatalytic process seems essential for the establishment of the kind of feedback between the two processes necessary for the maintenance of sustained oscillations. A simplified version of the FKN mechanism, called the "Oregonator", can be used as a basis for a computer simulation of the BZ reaction[22]. A numerical solution of the rate equations describing this mechanism shows many features of the reaction with initially added bromide; namely, initial decrease in bromide concentration, followed by a quiescent period, followed by sustained oscillations.

Although the BL reaction contains no organic compounds, the list of reactions to be considered is lengthy and complicated. For example, O_2 has to be considered as aquated (i.e., dissolved) reaction intermediate, and as gaseous product. Nevertheless, the principles of the FKN mechanism have been applied to the BL reaction with some success[23], though a number of questions still remain about certain details of the mechanism[24]. The mechanism contains a "non-radical" process,

Table 1. The FKN mechanism for the BZ reaction

Reaction	
Process (A)	
$Br^- + BrO_3^- + 2H^+ \rightarrow HBrO_2 + HOBr$	(R 3)
$Br^- + HBrO_2 + H^+ \rightarrow 2HOBr$	(R 2)
$3(Br^- + HOBr + H^+ \rightarrow Br_2 + H_2O)$	(R 1)
$3(Br_2 + CH_2(COOH)_2 \rightarrow BrCH(COOH)_2 + Br^- + H^+)$	(R 8)
$2Br^- + BrO_3^- + 3H^+ + 3CH_2(COOH)_2 \rightarrow 3BrCH(COOH)_2 + 3H_2O$	(A)
Process (B)	
$2(BrO_3^- + HBrO_2 + H^+ \rightarrow 2BrO_2 + H_2O)$	(R 5)
$4(Ce(III) + BrO_2 + H^+ \rightarrow Ce(IV) + HBrO_2)$	(R 6)
$2(2Ce(III) + BrO_3^- + HBrO_2 + 3H^+ \rightarrow 2Ce(IV) + H_2O + 2HBrO_2)$ (autocatalytic)	(G)
$2HBrO_2 \rightarrow HOBr + BrO_3^- + H^+$	(R 4)
followed by (R 1) and (R 8) to give:	
$BrO_3^- + 4Ce(III) + CH_2(COOH)_2 + 5H^+ \rightarrow 4Ce(IV) + BrCH(COOH)_2 + 3H_2O$	(B)
Also	
$4Ce(IV) + BrCH(COOH)_2 + 2H_2O \rightarrow 4Ce(III) + HCOOH + 2CO_2 + 5H^+ + Br^-$	(R 10)
Net Reaction (A) + 2 (B) + 2 (R 10)	
$3BrO_3^- + 3H^+ + 5CH_2(COOH)_2 \rightarrow 3BrCH(COOH)_2 + 2HCOOH + 4CO_2 + 5H_2O$	

$$2\,IO_3^- + 5\,H_2O_2 + 2\,H^+ \rightarrow I_2 + 5\,O_2(g.) + 6\,H_2O \quad \text{(II)}$$

a "radical" process

$$HIO_2 + 4\,H_2O_2 + I_2 \rightarrow IO_3^- + 2\,O_2(aq.) + 2\,I^- + 3\,H^+ + 3\,H_2O \quad \text{(III)}$$

and an autocatalytic species, HIO_2, which allows for the strong feedback that appears to be an integral part of sustained oscillations.

The most visually impressive oscillator is the BR reaction, because of the intense color developed by the starch indicator during phases of high iodine production. There have been two successful models which reproduce many characteristics of the temporal behavior of this system, including the oscillations[25, 26)]. The essential feature of these models is the presence of two separate processes, and the autocatalytic production of HIO_2, analogous to that of $HBrO_2$ in the BZ mechanism and HIO_2 in the BL mechanism.

III. Criteria for Oscillation

What conclusions can be drawn from these and other experimental and theoretical studies on chemical oscillators? Although it is not possible to specify a sufficient set of criteria that must be met to produce oscillations, some conditions seem to be necessary or at least helpful for oscillations to occur.

Since equilibrium is a stable state, reaction systems at, or close to, equilibrium will not oscillate[27)]. Therefore, one necessary condition for oscillation is that the system be *far from equilibrium*. We are, of necessity, dealing with open, irreversible systems. Second, and equally important, is the existence of a *feedback mechanism*. Oscillation or periodicity implies a return to some initial state. That this repetitive behavior would imply a particular kind of feedback, however, is not at all obvious.

The earliest hypothetical mechanism to yield an oscillatory solution to the ordinary differential equations of chemical kinetics was proposed in 1920 by Lotka (Table 2)[28)]. To visualize this mechanism, think of Y as wolves and X as cows, and a starting point of high X and low Y, together with an essentially unlimited supply of A, grass, which keeps the system open. Then notice two interesting features of this, for its day, avant-garde mechanism. First, as symbols of a predator-prey relationship, X and Y are coupled, and appear together in the *same* equation. Second, both equations L1) and L2) are autocatalytic. Within a single process, e.g., the rise and fall of the wolf population, autocatalysis is itself a feedback mechanism – one which produces exponential growth. Consider how autocatalysis affects one process in the mechanism, the Y (wolf population) cycle. Suppose Y is initially low and X high; autocatalytic growth leads to a rapid surge in the Y population with associated precipitous decrease in X. With little X to sustain it, Y peaks at some maximum value, then declines as reaction (L3) takes over. Since the system is open, X is regenerated so the fall in Y halts as high X values are again approached, and the system returns to its initial state of low Y and high X.

Autocatalysis is not the only mechanism of feedback which may lead to oscillations in chemical systems. Induced reactions and inhibition are also forms of feedback. These and other such mechanisms are discussed by Tyson[29)].

Table 2. The Lotka mechanism for an oscillating reaction[a]

$A + X \xrightarrow{k_1} 2X$	(L1)
$X + Y \xrightarrow{k_2} 2Y$	(L2)
$Y \xrightarrow{k_3} P$	(L3)

[a] Component is maintained constant

In addition to openness and feedback, a third condition is often found in conjunction with oscillation, though it has not yet been proven to be necessary: the existence of multiple stationary states. Most chemists are aware of the steady state condition, wherein the rate of change of the concentration of an intermediate can be equated to zero. Thus, the concentration of this intermediate can be obtained by solving an algebraic, rather than a differential equation. The familiar solution to this equation is a *single* concentration for a given set of initial conditions. [The reader may be familiar with a mechanism like $A + B \underset{k_{-1}}{\overset{k_1}{\rightleftharpoons}} X \xrightarrow{k_2} C$, and its steady-state solution for the concentration of X; $[X]_{SS} = k_1[A][B]/(k_{-1} + k_2)$.] This steady state is stable, although transitory, since perturbations of the steady-state concentration will decay back to the same concentration. Systems with autocatalytic steps do not always behave this way. Sufficiently large perturbations of their steady-states may lead to transitions to *different* steady-state concentrations. In other words, systems with autocatalytic steps may have multiple steady-states.

We will not attempt to review the literature on multiple stationary states here. However, two types of chemist independently became interested in multiple stationary states: chemical engineers and physical (especially biophysical) chemists. Schmitz reviews the early history of this subject from an engineering viewpoint[30]; Franck reviews the subject from a biophysical perspective[31]. More detailed theoretical descriptions may be found in papers by Nitzan et al.[32], and by Boissonade and DeKepper[33]. For our purposes, there are two important ideas connected with multiple stationary states: (1) reacting systems with these states show hysteresis, (2) for certain of these systems, changing a single parameter such as temperature or a reactant concentration near the transition point may destabilize the steady states so as to produce a limit cycle or oscillatory state. We illustrate these concepts through a simple model devised by Edelstein[34].

In this model (Table 3), substrate, A, is transformed to product, B, by an enzyme, E. The supply of A is large, ensuring far-from-equilibrium conditions. An intermediate, X, is produced autocatalytically, and degraded by the enzyme. (This feature of the model makes it unrealistic, as few autocatalytic processes arise this way.) The steady-state equation for X is cubic, and has three roots, or solutions for certain values of the parameters. One of the solutions is unstable; a real system cannot maintain a steady-state concentration, $[X]_{SS}$, with a value corresponding to this solution. Therefore, before $[X]_{SS}$ approaches such a value too closely, it jumps to a different value, corresponding to one of the stable solutions. This behavior leads to hysteresis, as shown in Fig. 1.

The existence of two stable steady states for the same values of the externally controlled constraints is known as "bistability". A mechanical analogy to such a situation is the double well potential shown in Fig. 2. (No potential function is known to exist for open chemical systems that would play the role of the gravitational potential energy in

Table 3. The Edelstein model of multiple steady states and hysteresis[a]

$A + X \underset{k_{-1}}{\overset{k_1}{\rightleftharpoons}} 2X$	(E 1)
$X + E \underset{k_{-2}}{\overset{k_2}{\rightleftharpoons}} C$	(E 2)
$C \underset{k_{-3}}{\overset{k_3}{\rightleftharpoons}} E + B$	(E 3)
$[X]_{SS}^3 + (2 + [B]_0 - [A]_0)[X]_{SS}^2 + ([E]_0 - [A]_0(2 + [B]_0))[X]_{SS} - [B]_0[E]_0 = 0$	(E 4)[b]

[a] The subscripts "SS" and "0" represent steady state and initial concentrations, respectively

[b] In deriving this equation all kinetic rate constants have been set equal to one

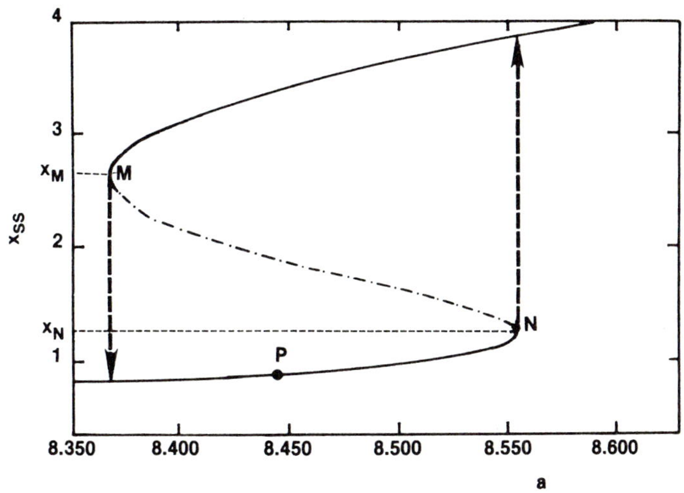

Fig. 1. Bistability and hysteresis in the Edelstein model[34]. In this figure x_{SS} ($[X]_{SS}$) is plotted as a function of a ([A]) for constant total enzyme concentration, $[E]_t = 30$, and $[B] = 0.2$. The upper and lower values (flow branch and thermodynamic branches, respectively) of x_{SS} are stable. The intermediate branch (*dashed line*) is unstable. Suppose one starts with a solution of composition P, on the thermodynamic branch, and then proceeds to increase a. Stable solutions of x_{SS} with increasing values result, until point N is reached. At this point an increase in a results in a transition to the flow branch. Decreasing a causes x_{SS} to decrease along the flow branch, until point M is reached, whereupon an abrupt transition back to the thermodynamic branch occurs. Increases in a bring x_{SS} back to P, completing a hysteresis loop

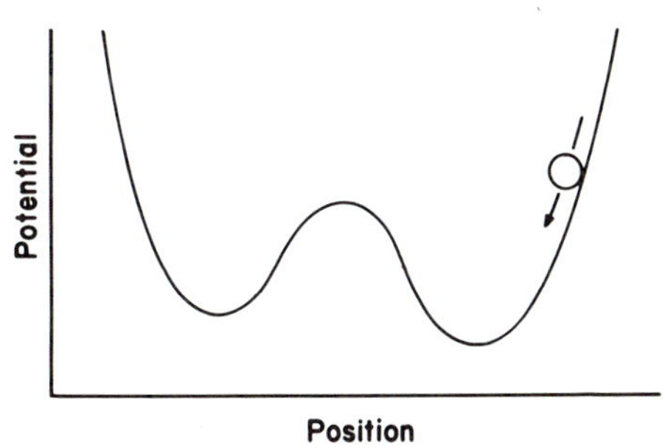

Fig. 2. A ball in a double well potential

the mechanical system.) Just as a ball in the well shown in Fig. 2 will oscillate between the two wells given a sufficient initial impulse, so one might imagine that a bistable chemical system, appropriately manipulated, might jump periodically between compositions close to those of the two steady state branches. While the analogy is far from exact, and the "appropriate manipulation" of the chemical system is much more complex than that of the mechanical one, the notion that bistability and oscillation are closely related has proved to be a fruitful one.

IV. Designing a Chemical Oscillator

The essence of our approach to designing new oscillators is: choose autocatalytic systems, keep them far from equilibrium, and look for multiple stationary states. Carrying out this program requires specialized experimental and theoretical "tools".

Obviously, to find a specific set of conditions for the onset of a concentration jump between stable states, or for the occurrence of oscillations, for that matter, would require the testing of a huge number of reaction mixtures under conditions maintained far from equilibrium. Since such an experiment is not easily accomplished in ordinary glassware, we conducted our studies of oscillatory reactions in continuous flow reactors patterned after those used at the Paul Pascal Research Center in Bordeaux, France[35].

The volume or space in which reaction occurs is called the reactor. Closed systems, for which matter is neither gained nor lost, are referred to in the engineering literature as "batch" reactors. An open, or "flow" reactor, which permits the flow of matter in and out of the system, allows for the continuous and convenient change of solution composition. Most importantly, the continuous flow of matter into and out of the flow reactor trivially solves the problem of maintaining the system far from equilibrium, while facilitating the detection and determination of the chemical properties of species in these states.

Thus, a continuously stirred tank reactor (CSTR) is the chief experimental tool that we have used for investigating the kinetics of reactions far from equilibrium[36]. The double-walled, pyrex reactor is designed to allow the simultaneous input of one to four solutions with a single outflow (in Fig. 3 the reactor depicted has three input channels). In the temperature controlled reactor one to three of the following properties can be determined: pH, absorbance of light, oxidation-reduction potential at a platinum electrode, potential of an ion-selective electrode (bromide or iodide). (In principle, temperature changes, electron spin resonance and/or conductivity changes could also be measured.) The variables at the disposal of the experimenter – the "external constraints" – are then the compositions of the solutions in the individual input reservoirs, the total flow rate, and the temperature of the thermostatic bath. Single or multiple addition of reagents, e.g., acid or base, to the reaction mixture is also possible.

Since few differential rate equations of interest can be integrated[37], mathematical analysis of oscillating reactions benefits from techniques developed for handling nonlinear equations. Using this approach, chemical engineers have examined consecutive, or coupled reactions from the standpoint of the type of behavior that such systems might be expected to exhibit[38]. This way of viewing a reaction is quite different from seeking an explicit integrated solution. However, as a means of introducing an alternative tech-

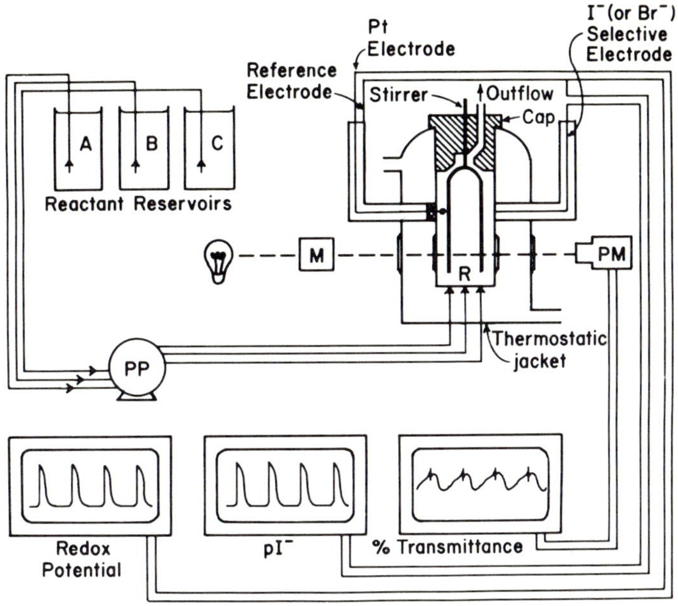

Fig. 3. Schematic diagram of a CSTR. In the configuration shown, up to three different solutions can be pumped (by the peristaltic pump, *PP*) into the reactor, *R*. The detectors shown in the diagram are: light absorption (*M*, monochromator; *PM*, photomultiplier), platinum (redox), and iodide (or bromide) selective electrodes. The reference electrode is the Hg/Hg_2SO_4 couple, in place of the usual calomel electrode, to avoid adventitious introduction of chloride into the reactor. In addition to these detectors, a thermocouple, or thermistor, and a pH electrode can be inserted into the reactor from above. The recordings of periodic behavior were taken from studies on the chlorite-iodide reaction

nique, let us return briefly, to one of the exact methods, the elimination of time as a variable[37].

If [A] and [P] are maintained constant, the Lotka mechanism (Table 2) can be described by two coupled, non-linear differential equations[28]

$$\frac{d[X]}{dt} = k_1[X] - k_2[X][Y] \tag{1}$$

$$\frac{d[Y]}{dt} = k_2[X][Y] - k_3[Y] \tag{2}$$

Division of (1) by (2) results in the elimination of time as a variable, and an equation in the two variables [X] and [Y] which can be integrated. The solution is

$$k_3 \log[X] - k_2[X] + k_1 \log[Y] - k_2[Y] = \text{constant} \tag{3}$$

A plot of [X] against [Y] is called a phase trajectory diagram or phase portrait[39]. Non-linear, but non-oscillatory reaction schemes have phase trajectories that are open curve segments,whose form depends on initial conditions and rate constants[37]. The phase trajectories of oscillatory reactions, like the Lotka mechanism, exhibit closed, non-intersecting trajectories[38, 39]. The exact appearance of the phase trajectory does not concern us here. What is of importance is that suitable plots of concentration from which time has been eliminated as an explicit variable can be diagnostic of the existence of such non-linear phenomena as multiple stationary states and oscillations.

The behavior of complex dynamical systems can be analyzed and represented in a number of ways. Figure 1 represents one such approach, a constraint-response plot. A constraint, in this case [A], is any variable which the experimenter can control directly. A response, $[X]_{SS}$ in this case, is a measurable property of the system which depends upon the constraint values. The constraints are the external variables, e.g., the temperature of the bath surrounding the reactor or the reservoir concentrations, while the responses are the internal variables, e.g., the temperature or concentration of species in the reactor. The phase trajectory diagram of Fig. 4 is one type of response-response plot. Obviously, in a complex system, there will be several constraints and responses subject to independent (or coupled) variation.

Let us consider the constraint-response plot of Fig. 1 a bit more closely. The (lower) set of solutions in Fig. 1 which extrapolates to vanishingly small [A] is termed the *equilibrium branch*. The (upper) set of solutions continuing to high [A] is termed the *flow branch*. The dotted portion of the sigmoidal curve is the *unstable branch*. To gain a better understanding of the conditions under which a system might be close to making jumps from one branch to another, we consider a different type of concentration diagram.

An alternative representation for information about the dynamics of complex systems is a constraint-constraint plot or *phase diagram*, in which values of the constraints for which particular types of behavior (single steady state, bistability, oscillation, etc.) occur are depicted. A number of analogies may be drawn between these non-equilibrium dynamical phase diagrams and those of equilibrium thermodynamics[32]. For our purposes, however, we need only take note of a type of phase diagram which plays a crucial role in the design of chemical oscillators, the "cross-shaped phase diagram" shown schematically in Fig. 5. The two lines represent the stability boundaries of the two branches of steady states. Where both states are stable we have bistability, where neither is stable, the system oscillates. In the immediate neighborhood of the cross point, the dynamics becomes extremely complex, and several exotic modes of behavior are possible[40].

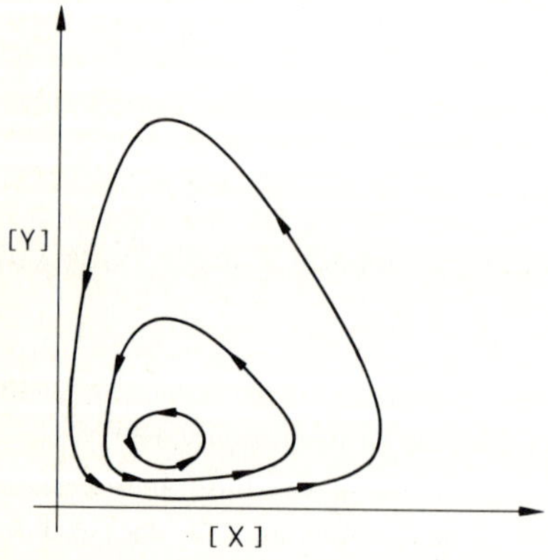

Fig. 4. Phase trajectory diagram for the Lotka mechanism of Table 2.

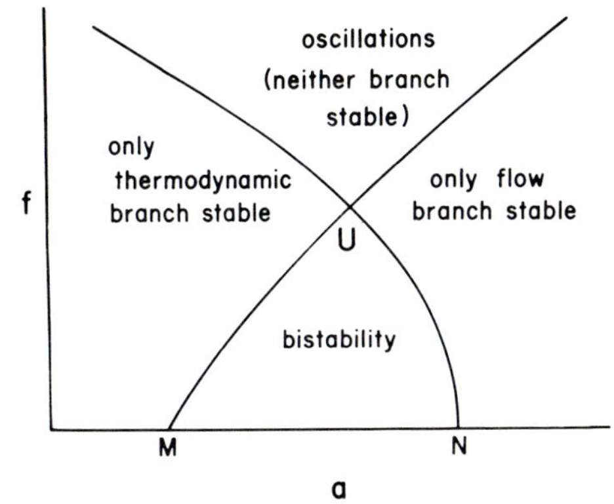

Fig. 5. Hypothetical cross-shaped phase diagram. Consider a system with a constraint-response diagram such as the one illustrated in Fig. 1. The steady state value of the response x_{SS} is determined by the constraint *a*. Suppose that another constraint *f* (e.g. another chemical species which reacts with x) is introduced which exerts a feedback effect on *a*. The value of *f* will modify the shape of the constraint-response curve, shifting the locations of points M and N, i.e., changing the regions of stability of the two branches. For an appropriately chosen feedback species[40], at high enough values of *f*, M will be found *to the right* of N and between these points, the only steady state will be the unstable one. For such values of *a* and *f*, the system will oscillate. At lower values of *f* we have essentially the same situation depicted in Fig. 1. This behavior is summarized in the cross-shaped phase diagram. Systems in the neighborhood of the cross point U are extremely sensitive to small changes in the constraint parameters

V. The Autocatalytic Reactions

We now turn to the selection of reactions exhibiting autocatalysis. Chemical reactions with abrupt, non-linear changes of concentrations have been known for some time. They are the spectacular "clock" reactions, first described by Landolt in the nineteenth century[41]. The key to understanding the sudden and predictable (you could set your watch by their occurrence, hence the name "clock") color changes was provided over 60 years ago by Eggert and Scharnow[42]. They analyzed the Landolt reaction (IV) (in the presence of excess iodate)

$$2\,IO_3^- + 5\,SO_3^{2-} + 2\,H^+ \rightarrow I_2 + 5\,SO_4^{2-} + H_2O \qquad \text{(IV)}$$

and pointed out that the iodide intermediate is produced autocatalytically according to the reaction

$$IO_3^- + 3\,SO_3^{2-} + 5\,I^- \rightarrow 6\,I^- + 3\,SO_4^{2-} \qquad \text{(V)}$$

Various reducing substrates such as arsenite, thiosulfate, ferrocyanide, etc. can substitute for sulfite, producing clock reactions in which autocatalysis can be demonstrated to occur. Our strategy was to combine two autocatalytic reactions having a species in common to afford strongly coupled feedback between the two subsystems.

The two systems we combined are the arsenite-iodate and the chlorite-iodide reactions. We first describe the arsenite subsystem.

In the presence of excess iodate, the net stoichiometry of this reaction is

$$2IO_3^- + 5H_3AsO_3 + 2H^+ = I_2 + 5H_3AsO_4 + H_2O \tag{VI}$$

Eggert and Scharnow identified three major component processes making up reaction (VI)[42].

$$IO_3^- + 3H_3AsO_3 = I^- + 3H_3AsO_4 \tag{VI A}$$

$$IO_3^- + 5I^- + 6H^+ = 3I_2 + 3H_2O \tag{VI B}$$

$$I_2 + H_3AsO_3 + H_2O = 2I^- + H_3AsO_4 + 2H^+ \tag{VI C}$$

Process (VI A) is the net reaction in excess arsenite ($[H_3AsO_3]_0/[IO_3^-]_0 > 3$); it is equivalent to (VI B) + 3 (VI C). Process (VI B), the Dushman reaction, is normally rate determining. Therefore, the net process (VI A) is autocatalytic in $[I^-]$, which causes a dramatic color change to arise at the stoichiometric point due to the sudden appearance of I_2. Bognár and Sárosi exploited this fact to devise a chronometric technique for the determination of traces of iodide[43]. Process (VI C), the Roebuck reaction[44], has played an important role in the development of chemical kinetics and teaching of physical chemistry, as it was the first multi-step reaction for which it was shown that the quotient of the independently determined forward and reverse rate laws equals the equilibrium quotient obtained from the law of mass action.

Reliable rate laws have been determined for processes (VI) (A)[42], (B)[43], and (C)[45], and are collected in Table 4. A resolution of this reaction into its component elementary steps is not required for an understanding of how the oscillator is developed. What is important is that we identify the autocatalytic process, which is (VI D) = (VI B) + 3 (VI C).

$$IO_3^- + 3H_3AsO_3 + 5I^- = 6I^- + 3H_3AsO_4 \tag{VI D}$$

(Note that (VI D) is equal to (VI A).) The reaction is initiated by (VI A); as soon as iodide builds up, however, processes (VI B) and (VI C) dominate the dynamics.

The dynamics of the iodate-arsenite reaction in a CSTR have been the object of two studies[46, 47]. Both studies find bistability in the oxidation of arsenite by iodate in acidic media (pH $\sim$ 2). At an apropriate flow rate the concentration of the autocatalytically produced intermediate, I^-, may take either of two stable steady-state values. At low flow rates there occurs a state characterized by relatively high values of $[I^-]$. This state, designated SSI, is a continuation of the thermodynamic or equilibrium branch of steady-state solutions. At flow rates higher with respect to the rates of the chemical reactions in the reactor, the steady-state values of $[I^-]$ are relatively low; this state is an extension of the flow branch of solutions, and is designated SSII. (A similar range of SS concentrations holds for I_2.) Either state SSI or SSII may exist for certain values of initial concentrations, $[IO_3^-]_0$ and $[H_3AsO_3]_0$, and flow rates, k_0. Transitions between these states may

Table 4. Component processes of the arsenite-iodate-chlorite oscillator and their rate laws[a]

Arsenite-Iodate

Overall reaction, $[H_3AsO_3]_0/[IO_3^-]_0 > 3:1$

$5\,I^- + IO_3^- + 6\,H^+ = 3\,I_2 + 3\,H_2O$ (A)

$3\,(H_3AsO_3 + I_2 + H_2O = H_3AsO_4 + 2\,I + 2\,H^+)$ (B)

$IO_3^- + 3\,H_3AsO_3 + 5\,I^- = 6\,I^- + 3\,H_3AsO_4$ (autocatalytic) (1′)

$IO_3^- + 3\,H_3AsO_3 = 3\,H_3AsO_4 + I^-$ (1)

$(1') = (1) = (A) + 3\,(B)$

Rate laws:

$$r_A = -d[IO_3^-]/dt = (k_{A1} + k_{A2}[I^-])[H^+]^2[I^-][IO_3^-]$$

$$r_B = -d[H_3AsO_3]/dt = k_B\frac{[H_3AsO_3][I_2]}{[H^+][I^-]}$$

Overall reaction, $[IO_3^-]_0/[H_3AsO_3]_0 > 1:3$

$5\,I^- + IO_3^- + 6\,H^+ = 3\,I_2 + 3\,H_2O$ (A)

$H_3AsO_3 + I_2 + H_2O = 3\,H_3AsO_4 + 2\,I^- + 2\,H^+$ (B)

$IO_3^- + 3\,H_3AsO_3 = I^- + 3\,H_3AsO_4$ (C)

$2\,IO_3^- + 5\,H_3AsO_3 + 2\,H^+ = I_2 + 5\,H_3AsO_4 + H_2O$ (2)

$(2) = 2\,(A) + 5\,(B) = (A) + 2\,(B) + (C)$

Rate laws: in addition to r_A, r_B

$$r_C = -d[H_3AsO_3]/dt = k_C[H_3AsO_3][IO_3^-]$$

Chlorite-Iodide

Overall reaction:

$ClO_2^- + 4\,I^- + 4\,H^+ \rightarrow Cl^- + 2\,I_2 + 2\,H_2O$ (rate determining) (D)

$5\,ClO_2^- + 2\,I_2 + 2\,H_2O \rightarrow 5\,Cl^- + 4\,IO_3^- + 4\,H^+$ (fast) (E)

$5\,I^- + IO_3^- + 6\,H^+ = 3\,I_2 + 3\,H_2O$ (slow) (A)

$3\,ClO_2^- + 2\,I^- \rightarrow 2\,IO_3^- + 3\,Cl^-$ (3)

$(3) = (1/2)[(D) + (E)] = (1/5)[3\,(E) + 2\,(A)]$

Rate law: in addition to r_A

$$r_D = -d[ClO_2^-]/dt = \left\{k_{D1}[H^+][I^-] + k_{D2}[I_2] + k_{D3}\frac{[I_2]}{[I^-]}\right\}[ClO_2^-]$$

$$r_E = -(1/2)d[I_2]/dt = \left\{\left(k_{E1} + \frac{k_{E2}}{[H^+]}\right)[ClO_2^-] + k_{E3}\right\}[I_2]$$

---

[a] Subscript zero represents initial concentration

be induced by suitable transient perturbations of sufficiently large amplitude, as shown in Fig. 6.

These experiments can be modeled by utilizing processes (VI) (A), (B) and (C), and the associated rate constants. A computer simulation of an experiment such as that shown in Fig. 6 is a stringent test for the validity of a set of rate constants.

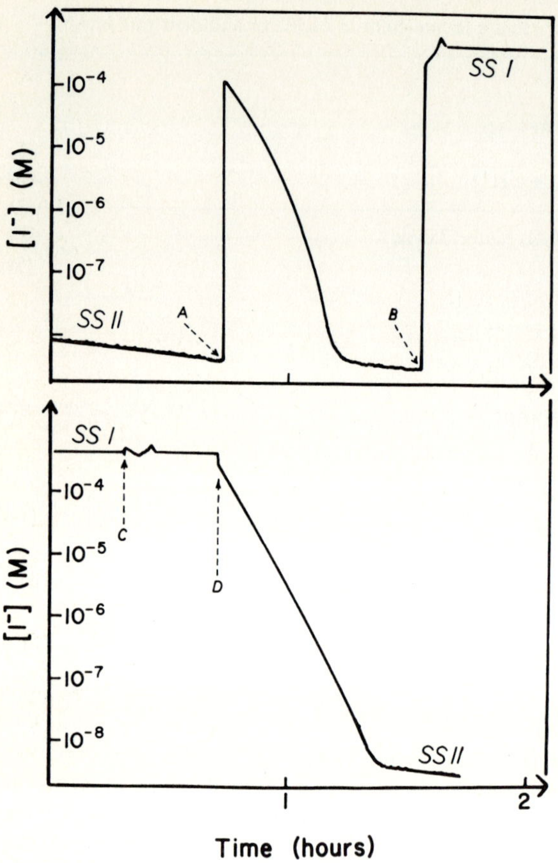

Fig. 6. Transition by perturbation between steady states with $[IO_3^-]_0 = 1.33 \times 10^{-3}$ M, $[H_3AsO_3]_0 = 2 \times 10^{-3}$ M, and $1/\tau = 2.2 \times 10^{-3}\,s^{-1}$, where τ = reactor residence time, T = 25 °C. At points A and B SSII is perturbed by sudden injections of KI into the CSTR to produce initial increases in $[I^-]$ of 1.3×10^{-4} and 3×10^{-4} M, respectively. At points C and D, SSI is perturbed by rapid additions of NaOH inducing sudden increases in $[OH^-]$ of 3×10^{-2} and 12×10^{-2} M, respectively

A more detailed mechanistic analysis of the arsenite-iodate reaction should give good agreement with several different types of experiments. These experiments would include: the Dushman reaction, the (batch) clock reaction, bistability in a CSTR, the Roebuck reaction, and spatial waves (see below). Clearly, a large number of reactions would have to be considered. After such a set is found, quantitative agreement requires the knowledge of many, perhaps all, of the rate constants for these elementary steps. Although such a full description of the arsenite-iodate reaction does not yet exist, a good start towards such a formulation has been made. As an indication of progress towards this end, plausible sets of reactions constituting a mechanism of the Dushman reaction (Table 5) and the arsenite-iodate reaction (Table 6) are given.

Before considering the chlorite reaction, we mention one further aspect of the arsenite-iodate reaction, its spatial inhomogeneity. Epik and Shub first reported chemi-

Table 5. Some mechanistic steps in the Dushman reaction

$5\,I^- + IO_3^- + 6\,H^+ = 3\,I_2 + 3\,H_2O$	(A)
Key Intermediate[a]: $H_2I_2O_3$	
$2\,H^+ + IO_3^- + I^- = H_2I_2O_3$	(1)
$H_2I_2O_3 + H^+ + I^- = 3\,IOH$	(2)
$H_2I_2O_3 = IOH + HIO_2$	(3)
$HIO_2 + I^- = 2\,IOH$	(4)
$IOH + H^+ + I^- = I_2 + H_2O$	(5)
$(A) = \underbrace{(1) + (2) + 3(5)}_{k_{A2}\text{-term}} = \underbrace{(1) + (3) + (4) + 3(5)}_{k_{A1}\text{-term}}$	

[a] This species may equally well be represented as I_2O_2, or $IO_2^+ \ldots I^-$, where IO_2^+ arises from $HIO_3 \rightleftarrows IO_2^+ + OH^-$

Table 6. Some mechanistic steps in the arsenite-iodate reaction

$IO_3^- + 3\,H_3AsO_3 = 3\,H_3AsO_4 + I^-$	(F)
$5\,I^- + IO_3^- + 6\,H^+ = 3\,I_2 + 3\,H_2O$	(A)
$IOH + I^- + H^+ = I_2 + H_2O$	(5)
$H_3AsO_3 + IO_3^- + H^+ \rightarrow H_3AsO_4 + HIO_2$	(6)
$H_3AsO_3 + HIO_2 \rightarrow H_3AsO_4 + IOH$	(7)
$H_3AsO_3 + I_2 + H_2O \rightarrow H_3AsO_4 + 2\,I^- + 2\,H^+$	(8)[a]
$(F) = (5) + (6) + (7) + (8) = (1/2)[(A) + (5) + (6) + (7) + 4(8)]$	

[a] Reaction between arsenite and hypoiodous acid (IOH) was ruled out by Liebhafsky [J. Phys. Chem. *34*, 1648 (1931)] on the basis of the best available evidence at the time. A recent reevaluation of the pertinent rate data[46], indicates that this reaction may occur. In that case, reactions (5) and (8) are replaced by $H_3AsO_3 + IOH \rightarrow H_3AsO_4 + I^- + H^+$ (8′) and $F = (1/2)[(A) + (6) + (7) + 4(8')]$

cal waves in the iodate oxidation of arsenite over 25 years ago[48]. This finding was subsequently confirmed and reinvestigated[49]. Waves are initiated in a narrow tube after an induction period has elapsed. Alternatively, waves can be initiated in thin layer solutions by pulsing a negatively biased Pt electrode. The propagation of the wave was studied as a function of concentration and pH. It was found that the characteristics of wave propagation could be simulated by a mechanism consisting of processes (VI) (B) and (C), together with diffusion terms. Thus, the essential autocatalytic nature of the reaction can be used to explain the existence of chemical waves in this system.

Like the arsenite-iodate reaction, chlorite-iodide is a clock reaction[50, 51]; however, it is more complex. This reaction shows a dramatic rise in the intensity of the brown color of iodine, followed by an even more abrupt fade-out. This behavior resembles that of the ferrous-nitrate clock, where the color is due to formation of the $FeNO^{2+}$ complex[52].

In the presence of excess ClO_2^-, the net reaction is

$$3\,ClO_2^- + 2\,I^- = 2\,IO_3^- + 3\,Cl^- \qquad \text{(VII)}$$

The first phase of the reaction (production of I_2):

$$ClO_2^- + 4I^- + 4H^+ = Cl^- + 2I_2 + 2H_2O \tag{VII A}$$

is autocatalytic in I_2. Further oxidation by chlorite is given by

$$5ClO_2^- + 2I_2 + 2H_2O = 5Cl^- + 4IO_3^- + 4H^+ \tag{VII B}$$

In these acidic media, with both iodate and iodide present, the Dushman reaction (VI B) will compete significantly with (VII A), thereby inhibiting (VII B). The net reaction is given by the relation (VII) = 1/2[(VII A) + (VII B)].

Rate laws for the two phases – growth and decay of iodine – of the chlorite-iodide reaction have been determined[50, 51, 53], and are collected in Table 4. For the purposes of

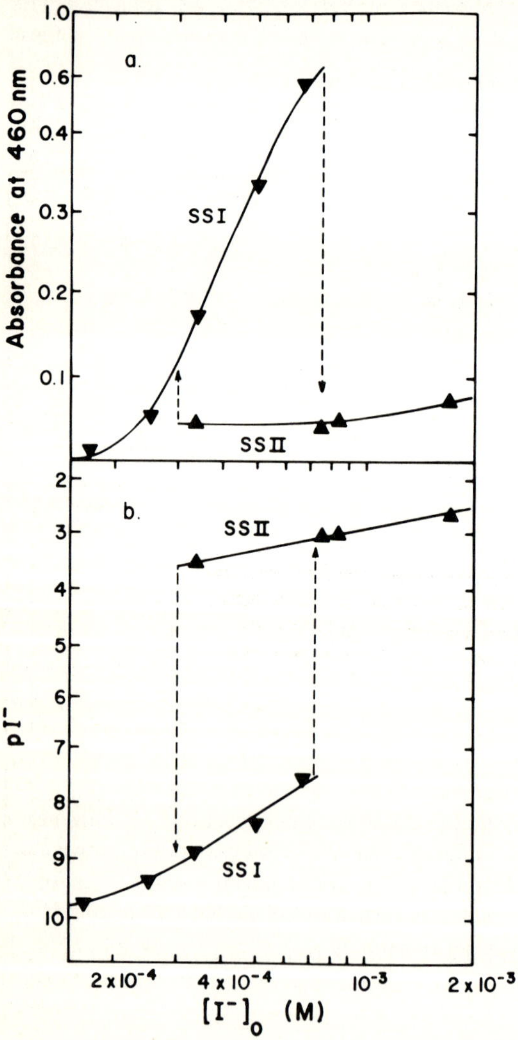

Fig. 7 a, b. Steady state hysteresis phenomena in the absorbance at 460 nm and the iodide concentration in the CSTR as a function of $[I^-]_0$, with $[ClO_2^-]_0 = 2.5 \times 10^{-4}$ M, pH 3.35, $1/\tau = 5.4 \times 10^{-3}\ s^{-1}$, and T = 25 °C, where τ = reactor residence time. *Dashed arrows* indicate spontaneous transitions between states

modeling the reaction on a time scale roughly equal to the period of oscillation, approximately five seconds to five minutes, reaction (VII B) is always complete. The hysteresis exhibited by the chlorite-iodide system is shown in Fig. 7. At low iodide flow ($[I^-]_0$), the system possesses only the single low iodide steady-state SSI, which we identify with the thermodynamic branch[54]. As $[I^-]_0$ is increased the system remains in SSI and $[I^-]$ in the reactor gradually increases until, when $[I^-]_0$ reaches 7.5×10^{-4} M, an abrupt transition to the high iodide steady-state SSII takes place. This state, the only one stable at higher iodide flows, is the continuation of the flow branch. If, starting from SSII, we decrease $[I^-]_0$, a smooth decrease in $[I^-]$ occurs until at an iodide flow of 2.5×10^{-4} M, the system vaults back to SSI with a concomitant drop in $[I^-]$ by four orders of magnitude.

The very narrow region of the constraint space where oscillations could occur in this system was not found until after the chlorite-iodate-arsenite oscillator had been discovered. Its occurrence is shown in the cross-shaped diagram for chlorite-iodide (Fig. 8.)

A complete mechanism for the chlorite-iodide system entails many steps. In addition to those consistent with the rate laws for processes (VII) (A) and (B), the elementary steps of the Dushman reaction need to be included. A set of relevant steps for these processes in collected in Table 7.

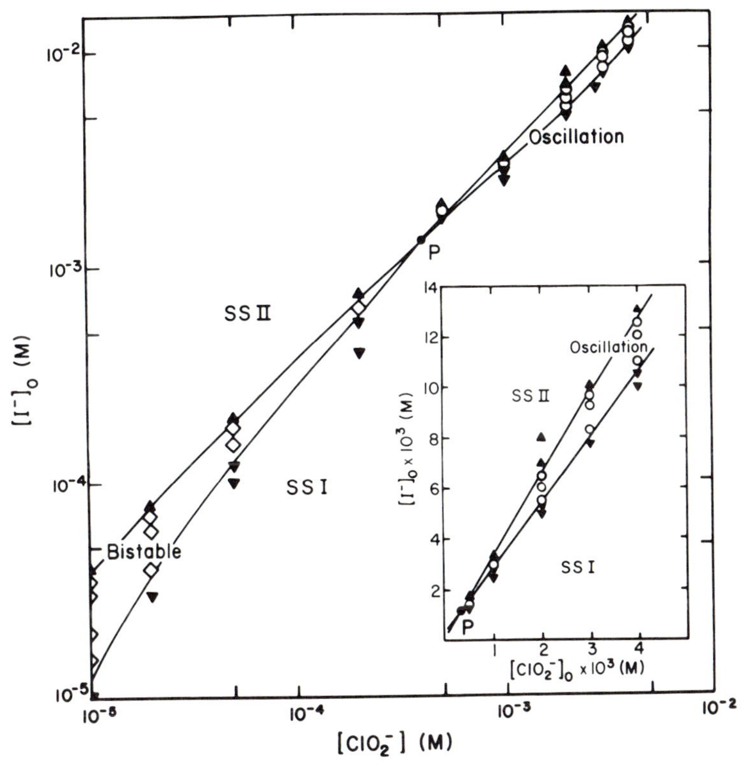

Fig. 8. Log-log plot of a section of the phase diagram in the $[ClO_2^-]_0$-$[I^-]_0$ plane with pH 2.04, $1/\tau = 1.1 \times 10^{-3}\ s^{-1}$, and T = 25 °C, where τ = reactor residence time. ▲, SSI; ▼, SSII, ◇, bistable; ○, oscillatory. Inset shows the same data on a linear scale. *P* indicates cross point

Table 7. Some mechanistic steps in the chlorite-iodide reaction

Reaction	
Process (D)[a]: $ClO_2^- + 4I^- + 4H^+ \rightarrow Cl^- + 2I_2 + 2H_2O$	
$ClO_2^- + I^- + H^+ \rightarrow ClO^- + IOH$ (slow, k_{D1})	(1)
$ClO^- + H^+ \rightarrow ClOH$	(1′)
$ClOH + I^- \rightarrow Cl^- + IOH$	(2)
$I^- + H^+ + IOH \rightarrow I_2 + H_2O$	(3)
$IOH + ClO_2 + H^+ \rightarrow HIO_2 + HOCl$ } autocatalytic in IOH.	(4)
$HIO_2 + I^- + H^+ \rightarrow 2IOH$ } k_{D2}	(5)
(D) = (1) + (1′) + (2) + 2(3) = (2) + (4) + (5) + 2(3)	
Process (E): $5ClO_2^- + 2I_2 + 2H_2O \rightarrow 5Cl^- + 4IO_3^- + 4H^+$	
$ClO_2^- + I_2 \rightarrow IClO_2 + I^-$, k_{E1}	(6)
$I_2 + H_2O \rightarrow I_2OH^- + H^+$	(7)
$ClO_2^- + I_2OH^- + H^+ \rightarrow IClO_2 + I^- + H_2O$, k_{E2}	(8)
$I_2 + H_2O = IOH_2^+ + I^-$	(9)
$ClO_2^- + IOH_2^+ \rightarrow IClO_2 + H_2O$, k_{E3}	(10)
$IClO_2 + H_2O \rightarrow IO_3^- + Cl^- + 2H^+$	(11)
$I^- + 5IO_3^- + 6H^+ \rightarrow 3I_2 + 3H_2O$	(A)
(E) = 5[(6) + (11)] + (A) = 5[(7) + (8) + (11)] + (A)	
= 5[(9) + (10) + (11)] + (A)	

[a] Lower valent states of iodine and chlorine in the presence of one another should generate interhalogen compounds and ions such as ICl, I_2Cl^- etc. These species, and polyanions such as I_3^-, should be considered in a more detailed mechanism

VI. The First Systematically Designed Oscillator

Both the chlorite-iodide and arsenite-iodate reactions fulfill the criteria set forth above to serve as the basis for an oscillator. Each is autocatalytic and in a CSTR exhibits bistability. In addition, I_2, the product of the ClO_2^--I^- reaction is a key intermediate in the H_3AsO_3-IO_3^- system, while iodide, the product of the latter reaction is a reactant in the former. The combined reaction thus seems ideal as a candidate for a systematically designed oscillator and so it has proved to be[55]. The ensuing oscillations – the first chemical oscillations generated as the result of a deliberate design program – are shown in Fig. 9.

From the design point of view, it is of interest to ask which of the subsystems, chlorite-iodide or arsenite-iodide represents the "fundamental" bistable part of this system and which species in the terminology of the cross-shaped diagram approach provides the feedback. This question can be answered by two independent means. Happily, they both point in the same direction; to an answer which leads us to the design of more new oscillators. The cross-shaped diagram analysis[33] requires that relaxation to the steady-state branches of the fundamental bistable subsystem be rapid compared with the time required for the feedback reaction. Since the chlorite-iodide reaction in a CSTR shows far more rapid relaxation than does the arsenite-iodate reaction, we are led to designate ClO_2^--I^- as the fundamental system with arsenite as the feedback species. Alternatively, we may think of the oscillation as occurring essentially between the two bistable steady-states of the fundamental subsystem under the impetus of the feedback. Figure 9 shows

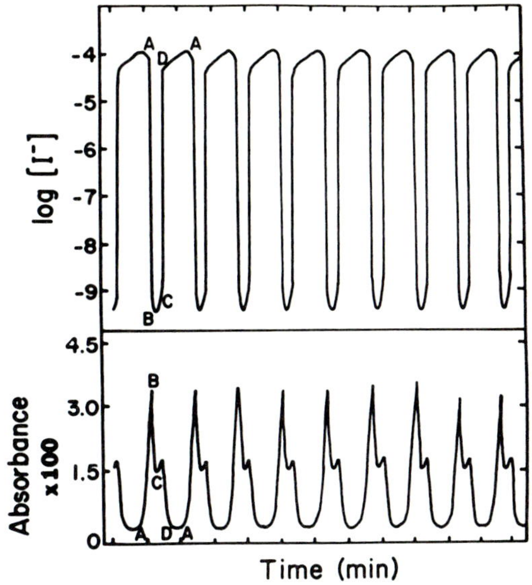

Fig. 9. Oscillations of the iodide concentration and absorption per cm path length at 460 nm for $[KIO_3]_0 = 24 \times 10^{-3}$ M, $[As_2O_3]_0 = 2 \times 10^{-3}$ M, $[NaClO_2] = 2 \times 10^{-3}$ M with $[Na_2SO_4]_0 = 0.1$ M, $[H_2SO_4]_0 = 0.01$ M, residence time = 400 s and T = 25 °C. Concentrations are given in the reactor after mixing but before any reaction takes place

that for most of the course of the reaction in the oscillating system $[I^-]$ is low when $[I_2]$ is high and *vice versa.* Comparing Figs. 7 and 10, which show the steady-state $[I^-]$ and $[I_2]$ for the two subsystems, we find that the observed behavior of the chlorite-iodate-arsenite system more closely resembles the steady-states of the ClO_2^--I^- subsystem than of the H_3AsO_3-IO_3^- reactor, for which steady-state values of $[I^-]$ and $[I_2]$ are simultaneously either high or low. The identification of the chlorite-iodide subsystem as the main oscillatory component, and of arsenite as the feedback species is an important step towards achieving a mechanistic understanding of the oscillator.

The development of an adequate mechanism for the BZ reaction required nearly 15 years from the discovery of oscillations in that system, and refinement of that mechanism is still under way[56]. It is a measure of the progress in the field of oscillating reactions that only 15 months after the design of the first chlorite oscillator, a mechanism for that system seems well within reach. Without setting forth a full mechanistic treatment, which is not yet available, we sketch here what we believe to be the key elements in the oscillation of the chlorite-iodate-arsenite oscillator and, by extension, several of the related systems to be discussed below. A partial mechanism for the prototype chlorite-iodide system will be presented in the following section.

Consider the oscillations shown in Fig. 9 and let us start our analysis at the point labeled A. We need to recall the following component processes

$$ClO_2^- + 4I^- + 4H^+ = Cl^- + 2I_2 + 2H_2O \qquad \text{(VII A)}$$
$$5ClO_2^- + 2I_2 + 2H_2O = 5Cl^- + 4IO_3^- + 4H^+ \qquad \text{(VII B)}$$
$$IO_3^- + 5I^- + 6H^+ = 3I_2 + 3H_2O \qquad \text{(VI B)}$$
$$I_2 + H_3AsO_3 + H_2O = 2I^- + H_3AsO_4 + 2H^+ \qquad \text{(VI C)}$$

At A, $[I_2]$ is low, $[I^-]$ high and nearly all the arsenite has been reduced to arsenite (this last observation is confirmed in Fig. 11 which shows a closely related oscillating reac-

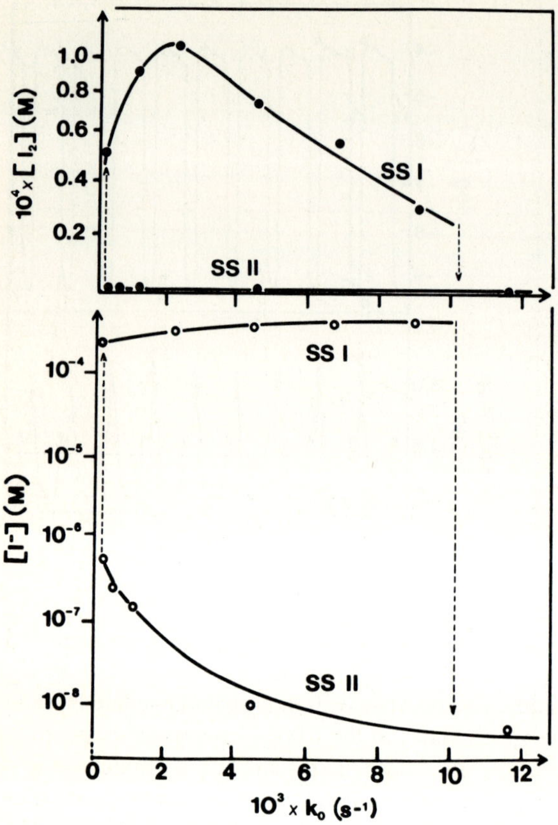

Fig. 10. Experimentally determined $[I_2]_{SS}$ and $[I^-]_{SS}$ as a function of flow rate with $[IO_3^-]_0 = 1.33 \times 10^{-3}$ M and $[H_3AsO_3]_0 = 2 \times 10^{-3}$ M. *Arrows* indicate transitions from one steady state to the other as the flow rate is varied. (At all k_0 $[I_2]_{SSII}$ is too small to measure; k_0 is reciprocal of the reactor residence time.)

tion[57] in which H_3AsO_3 has been replaced by another reductant and the redox potential as well as $[I^-]$ and absorbance are monitored).

Since there is no iodine availabe, process (VI C) is quiescent and I_2 begins to accumulate in processes (VII A) and (VI B), both of which destroy I^-. Since (VII A) is autocatalytic, this production of I_2 at the expense of I^- accelerates rapidly, bringing the system to the point labeled B, where $[I^-]$ has been almost totally depleted, which shuts down process (VII A). Processes (VII B) and (VI C) now begin to consume I_2 in a very rapid reaction, converting most of the arsenite to arsenate and regenerating some iodide. We are now at point C. Arsenite is replenished by the input flow (and arsenate flows out of the reactor) so that process (VI C) now provides a further route for the replenishment of I^-. Iodide accumulates slowly at first, but then the autocatalytic nature of the combined process (VI B) + (VI C) takes over, raising $[I^-]$ rapidly while bringing $[I_2]$ to a low pseudo-steady-state value at point D until the system returns to A and the cycle is ready to start over. The reason for the shoulder on the absorbance peak in Fig. 9 is obscure. It

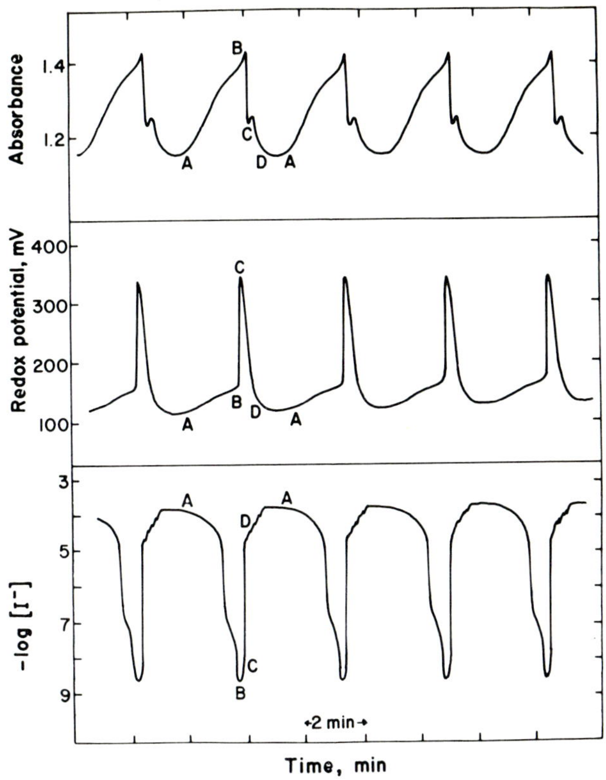

Fig. 11. Optical density at 460 nm, potential (compared with Hg/Hg_2SO_4 reference electrode) of platinum electrode, and $[I^-]$ determined by iodide sensitive electrode in a CSTR with input composition concentrations in the reactor after mixing, but before initiation of reactions: $[NaClO_2]_0$ = 0.002 M, $[KIO_3]_0$ = 0.01 M, $[K_4(Fe(CN)_6)]_0$ = 0.005 M. Temperature, 25 °C; pH, 2.06; residence time, 310 s. Points *A, B, C* and *D* correspond to those in Fig. 9

may represent a genuine secondary increase in $[I_2]$ or $[I_3^-]$) or it may result from a burst of ClO_2 (which also absorbs strongly at this wavelength) when processes (VII A) and (VII B) are relatively slow and $[H^+]$ is temporarily high as a result of [VII A] and [VI C].

VII. Chlorite Oscillators: An Overview

The chlorite-iodate-arsenite oscillator was the first oscillating reaction discovered which is based upon chlorite chemistry. The BZ reaction and its relatives are bromate oscillators, while the BL and Briggs-Rauscher oscillators are iodate systems. The initial chlorite oscillator was rapidly followed by a large family of related systems[58–60], which are summarized in Table 8. We note that while most of these systems contain an iodine species (I^-, I_2, IO_3^-) as well as the chlorite, at least two iodine-free chlorite oscillators exist.

Table 8. Chlorite Oscillators in a CSTR

Number	System	Special features	Ref.
1	ClO_2^--I^-	Bistability between stationary and oscillating states	54
2	ClO_2^--I^--IO_3^-		
3	ClO_2^--I^--malonic acid	Batch oscillation spatial wave patterns	59
4	ClO_2^--I^--$Cr_2O_7^{2-}$		58
5	ClO_2^--I^--MnO_4^-		58
6	ClO_2^--I^--BrO_3^-	Bistability between stationary and oscillating states	58
7	ClO_2^--IO_3^--H_3AsO_3	First chlorite oscillator discovered	55
8 a	ClO_2^--IO_3^--$Fe(CN)_6^{4-}$		57
8 b	ClO_2^--I_2-$Fe(CN)_6^{4-}$		58
9 a	ClO_2^--IO_3^--SO_3^{2-}		57
9 b	ClO_2^--I_2-SO_3^{2-}		58
10 a	ClO_2^--IO_3^--$S_2O_3^{2-}$	Batch oscillation	59
10 b	ClO_2^--I_2-$S_2O_3^{2-}$		58
11	ClO_2^--IO_3^--$CH_2O \cdot HSO_2$		57
12	ClO_2^--IO_3^--ascorbic acid		57
13	ClO_2^--IO_3^--I^--malonic acid	Batch oscillation	57
14	ClO_2^--I^--IO_3^--H_3AsO_3	Tristability	59
15	ClO_2^--$S_2O_3^{2-}$	First iodine-free chlorite oscillator	60
16	ClO_2^--Br^--BrO_3^-		[a]

[a] Orbán, M.; Epstein, I. R.: to be published

Some of these chlorite oscillators exhibit particularly interesting or exotic phenomena. Batch oscillations in the absence of flow may be obtained in the systems numbered 3, 10 a and 13, while the chlorite-iodide-malonic acid reaction gives rise to spatial wave patterns as well. These latter, which are strikingly similar to those observed in the BZ reaction[61)] are shown in Fig. 12. Addition of iodide to the original chlorite-iodate-arsenite oscillator produces a system with an extremely complex phase diagram[58)], shown in Fig. 13, which even contains a region of *tristability*, three possible stable steady-states for the same values of the constraints.

On studying the fundamental chlorite-iodide subsystem by itself[54)], we discovered that it, too, was an oscillator, though over a somewhat narrower range of flow rates and pH than the arsenite-containing system. The ClO_2^--I^- reaction also exhibits another type of bistability, known technically as subcritical Hopf bifurcation[62)], in which both a stable steady-state and a stable oscillatory state are possible for the same set of constraints. This behavior is illustrated in Fig. 14. Another exotic phenomenon, found in the chlorite-thiosulfate oscillator[60)], is that of complex oscillations, with several peaks per period. As Fig. 15 shows, as the flow rate in the CSTR is decreased, the system progresses from simple large amplitude oscillations through a series of complex periodic modes containing one large and as many as 16 small peaks per cycle until a state of single small amplitude oscillations is reached.

In an effort to produce a general scheme for understanding and categorizing chlorite-iodine oscillators, Orbán et al.[58)] have developed a set of "component stoichiometric

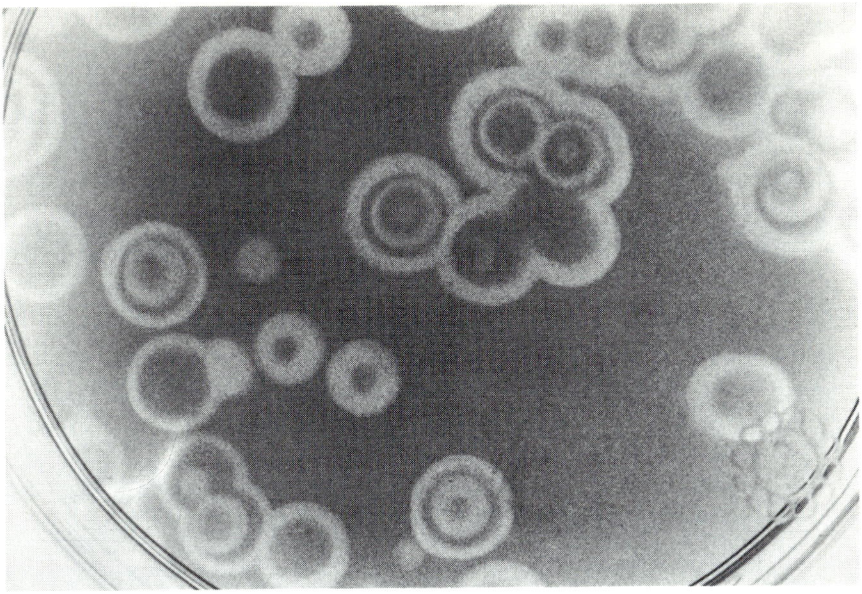

Fig. 12. Spatial wave pattern observed at 5 °C in a thin (2 mm) layer of reactive solution with initial composition $[CH_2(COOH)_2]$ = 0.0033 M, [NaI] = 0.09 M, $[NaClO_2]$ = 0.1 M, $[H_2SO_4]$ = 0.0056 M, and starch as indicator

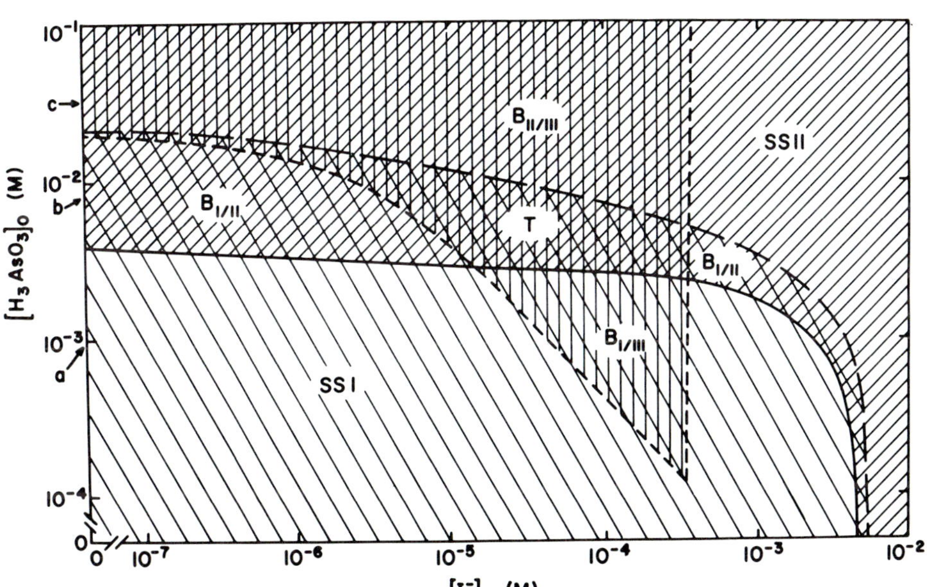

Fig. 13. Section of the phase diagram of the chlorite-iodie-iodate-arsenite system showing tristability in the $[I^-]_0$-$[H_3AsO_3]_0$ plane. ▧, SS I stable; ▨, SS II stable; ▥, SS III stable. $B_{i/j}$ indicates region of bistability of SS_i and SS_j. T indicates region of tristability. Fixed constraints: $[ClO_2^-]_0$ = 2.5×10^{-3} M; $[IO_3^-]_0 = 2.5 \times 10^{-2}$ M, pH = 3.35, $k_0 = 5.35 \times 10^{-3}\,s^{-1}$, T = 25 °C, k_0 is reciprocal of reactor residence time

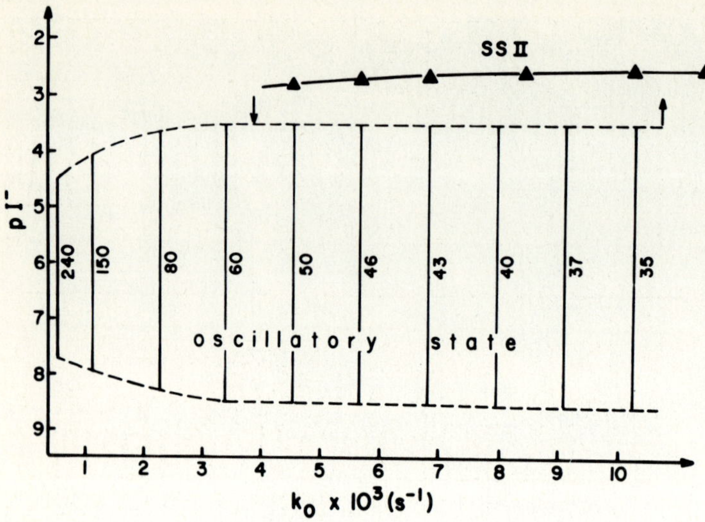

Fig. 14. Hysteresis in the transition between SSII and the oscillatory state as a function of flow rate k_0 with $[I^-]_0$ = 0.0065 M, $[ClO_2^-]$ = 0.002 M, pH 1.56, and T = 25 °C. Envelopes of vertical segments show upper and lower limits of pI^- in the oscillatory state. Numbers next to these segments indicate period of oscillation in seconds. *Arrows* indicate spontaneous transitions between states

processes" which may occur in these systems. Noyes had earlier presented such a grouping for bromate oscillators[63]. These processes are summarized in Table 9. Reactions (M1)–(M3) are the major processes (corresponding to our earlier reactions (VIB), VIIA) and (VIIb), respectively) taking place in systems containing chlorite and iodide, iodine and/or iodate, at least at pH's sufficiently high so that ClO_2 does not play a major role[64]. All three processes are strongly favored thermodynamically. Reactions (M4)–(M10), in which Red and Ox represent one-electron reducing or oxidizing substrates, respectively, account for the role of the additional species in these reactions. For two-electron reductants or oxidants, appropriate changes in the stoichiometries of these reactions should be made. A glance at Table 8 shows that chlorite oscillators can be constructed with either one- or two-electron substrates.

With this framework of component processes in mind, Orbán et al.[58] proposed the following preliminary classification of chlorite oscillators:

A) Chlorite-iodide. This is the fundamental or minimal chlorite-iodine oscillator in that it contains the minimal set of reactants necessary to generate processes (M1)–(M3).

A') Chlorite-iodide-iodate. While this system fits into either categories B) or C) below, it may also be considered a sort of fundamental oscillator which is generated (via M4, M8 or M9) by the systems of type B) or C).

B) Chlorite-iodide-oxidant. These systems are represented by oscillators 2, 4, 5 and 6 in Table 8. They are identified by the fact that they exhibit oscillations over a significantly broader range of constraints than the fundamental ClO_2^--I^- subsystem. The chief requirements on the substrate Ox are that (M4) be thermodynamically favorable ($E^0_{Ox/Red} > 0.54$ V) and rapid, (though not faster than (M2) + (M3)) and that

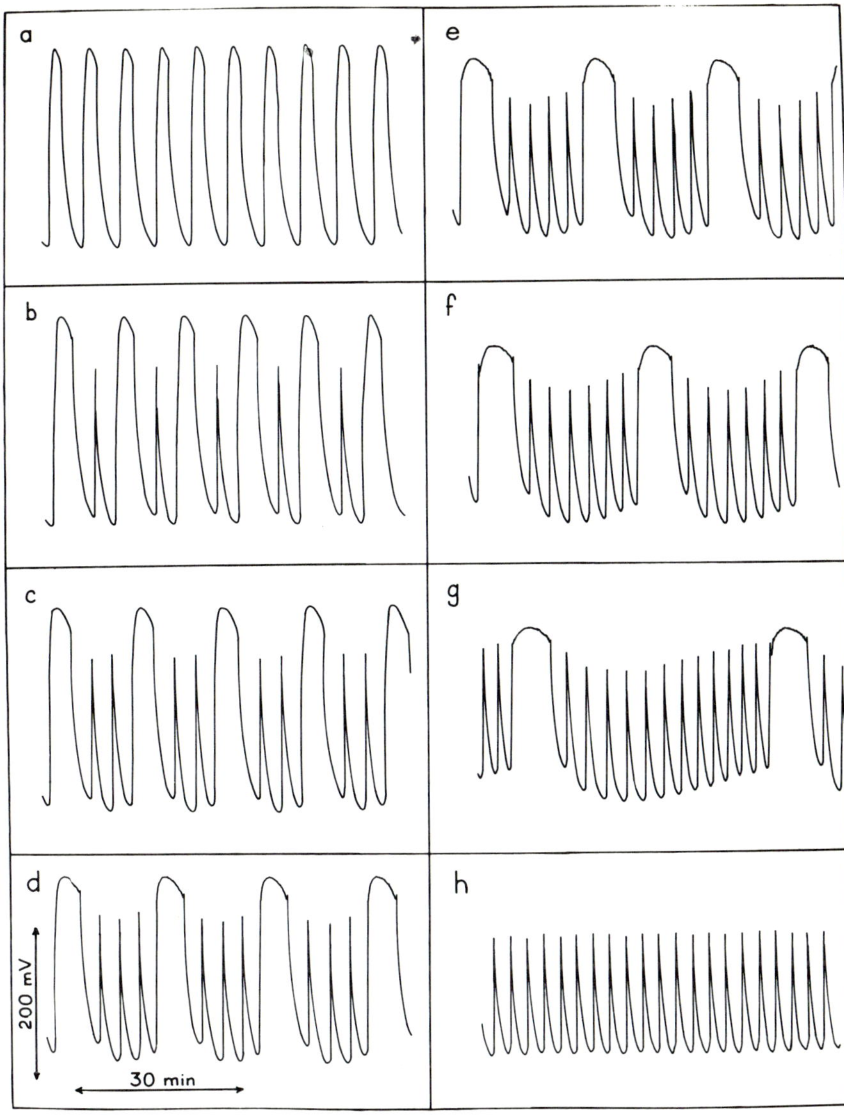

Fig. 15a–h. Complex periodic oscillations in the chlorite-thiosulfate system with $[ClO_2^-]_0$ = 5×10^{-4} M, $[S_2O_3^{2-}]_0 = 3 \times 10^{-4}$ M, pH 4, T = 25.0 °C. Residence times: **a** 5.9 min, **b** 9.5 min, **c** 10.8 min, **d** 13.5 min, **e** 15.8 min, **f** 20.6 min, **g** 26.3 min, **h** 47.2 min.

(M 7) be slow in some accessible range of pH and reactant fluxes. One oxidant which fails to enhance the range of constraints over which a chlorite-iodide system will oscillate is peroxydisulfate. In spite of its highly favorable reduction potential (E^0 = 2.01 V), $S_2O_8^{2-}$ reacts too slowly with I^- in reaction (M 4) to be a useful oscillatory substrate.

C) Chlorite-iodine species-reductant. Several subclasses may be distinguished within this category according to the oxidation number of the iodine species involved:

Table 9. Component processes in bromate and chlorite oscillators[a]

Bromate Oscillators	
$Br(V) + Br^- \rightarrow Br(III) + Br(I)$	(A 1)
$Br(III) + Br^- \rightarrow 2\,Br(I)$	(A 2)
$Br(V) + 2\,Br^- \rightarrow 3\,Br(I)$	(A)
$2x(Br(V) + Br(III) \rightarrow 2\,Br(IV))$	(B 1)
$4x(Br(IV) + Red \rightarrow Br(III) + Ox)$	(B 2)
$2\,Br(III) \rightarrow Br(V) + Br(I)$	(B 3)
$Br(V) + 4\,Red \rightarrow Br(I) + 4\,Ox$	(B)
$Br(I) + nOx \rightarrow Br^- + nRed + x$	(C)
$pBr(I) + qBr^- \rightarrow$ inert-Br	(D)
Chlorite Oscillators	
$I(V) + 5\,I^- \rightarrow 6\,I(0)$	(M 1)
$4\,I^- + Cl(III) \rightarrow 4\,I(0) + Cl^-$	(M 2)
$4\,I(0) + 5\,Cl(III) \rightarrow 4\,I(V) + 5\,Cl^-$	(M 3)
$Ox + I^- \rightarrow Red + I(0)$	(M 4)
$5\,Ox + I(0) \rightarrow 5\,Red + I(V)$	(M 5)
$Ox + Cl(III) \rightarrow Red + Cl(IV)$	(M 6)
$2\,Ox + Cl(III) \rightarrow 2\,Red + Cl(V)$	(M 7)
$I(V) + 6\,Red \rightarrow I^- + 6\,Ox$	(M 8)
$I(0) + Red \rightarrow I^- + Ox$	(M 9)
$Cl(III) + 4\,Red \rightarrow Cl^- + 4\,Ox$	(M 10)

[a] Roman numerals in parentheses refer to oxidation states of halogens, e.g. $I(0) = \frac{1}{2}\,I_2$, $Br(V) = BrO_3^-$, etc. Red and Ox are one equivalent reductants and oxidants, respectively

C′) Chlorite-iodate-reductant. Oscillators 2, 7, 8 a, 9 a, 10 a, 11 and 12 of Table 8 are of this type. The requirements for a successful substrate appear to be that (M 8) be thermodynamically favorable and relatively rapid while (M 10) is slow.

C″) Chlorite-iodine-reductant. These systems, which include systems 8 b, 9 b and 10 b of Table 8 appear to be only minor variants of type C′) in which (M 9) replaces (M 8).

C‴) Chlorite-iodide-reductant. The only known example of this type is the chlorite-iodide-malonic acid system, which is of special interest because it supports both batch oscillations and spatial wave patterns. The slow decomposition of iodinated malonic acid species apparently provides a long lasting, indirect flux of iodide (via (M 2) + (M 9)) in this system.

D) Iodine-free chlorite oscillators. In view of our almost total ignorance of how the chlorite-thiosulfate system functions, we place it for the moment in a class of its own, though further study may ultimately situate it in an expanded category A. The recently discovered chlorite-bromide-bromate oscillator may be analogous to the chlorite-iodide-iodate system of class A′) above, though one may view it alternatively as a bromate driven oscillator in which ClO_2^- plays the role of the metal catalyst.

While no complete mechanism has yet been developed which predicts oscillation in a chlorite oscillator from the integration of a set of rate equations derived from elementary

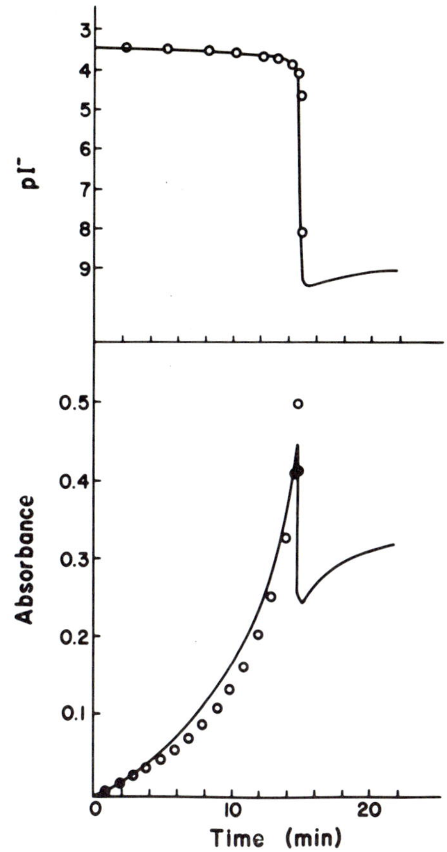

Fig. 16. Calculated (*circles*) and experimental (*lines*) absorbance at 460 nm (I_2 concentration) and iodide concentration for a batch reaction at 25 °C and pH 3.3 with initial concentrations $[I^-] = 4 \times 10^{-4}$ M, $[ClO_2^-] = 2.5 \times 10^{-4}$ M

steps, considerable progress has been made toward that goal. DeKepper et al.[47] and Papsin et al.[46] have shown that the bistable behavior of the arsenite-iodate subsystem can be simulated quite accurately by a relatively simple mechanism. Recent unpublished work in this laboratory has produced a preliminary mechanism consisting of some 20 elementary steps which provides almost quantitative agreement with the results of batch experiments on the chlorite-iodide, chlorite-iodine and iodide-iodate subsystems. A comparison between simulation and experiment is shown in Fig. 16. This mechanism involves such closed shell intermediate species, which were suggested by kinetics studies of the various subsystems[45, 50, 51, 53], as I_2O_2, $IClO_2$, IOH_2^+ and I_2OH^-. It does not, however, yet include such radical species as ClO_2, IO_2 or IO, which are likely to be essential for describing the oscillations under flow conditions.

VIII. Summary and Future Trends

The study of chemical oscillation has come a long way in the last 25 years, from an area of questionable validity populated by a handful of believers to a field of increasing importance in physical and inorganic chemistry. In a sense, the construction of a family of new oscillators by application of a systematic approach combined with some knowledge of inorganic reaction mechanisms constitutes the final proof that these systems are more than laboratory curiosities. To reiterate the algorithm which led to the discovery of the chlorite oscillators:

a) select an autocatalytic reaction,
b) run it in a flow reactor and find conditions under which it is bistable,
c) add a feedback species which exerts, at a relatively slow rate, a different perturbing effect on the two bistable branches.

While mechanistic understanding of even the earliest known oscillators is far from complete, it appears useful to summarize the known oscillating reactions by grouping them into classes which share common features. The most obvious such grouping results when one observes that all the known homogeneous oscillators contain an oxyhalogen ion; iodate, bromate or chlorite. We may thus identify iodate-, bromate- and chlorite-driven oscillators as our first major groupings, though there will be occasional ambiguity, as with the ClO_2^--BrO_3^--Br^- reaction. It is amusing to note that while chronologically the order of discovery is iodate, bromate, chlorite, the variety of systems found in each category lies in exactly the reverse order. The reason for this seeming paradox is largely historical in that iodate and bromate oscillators were first discovered in batch systems,

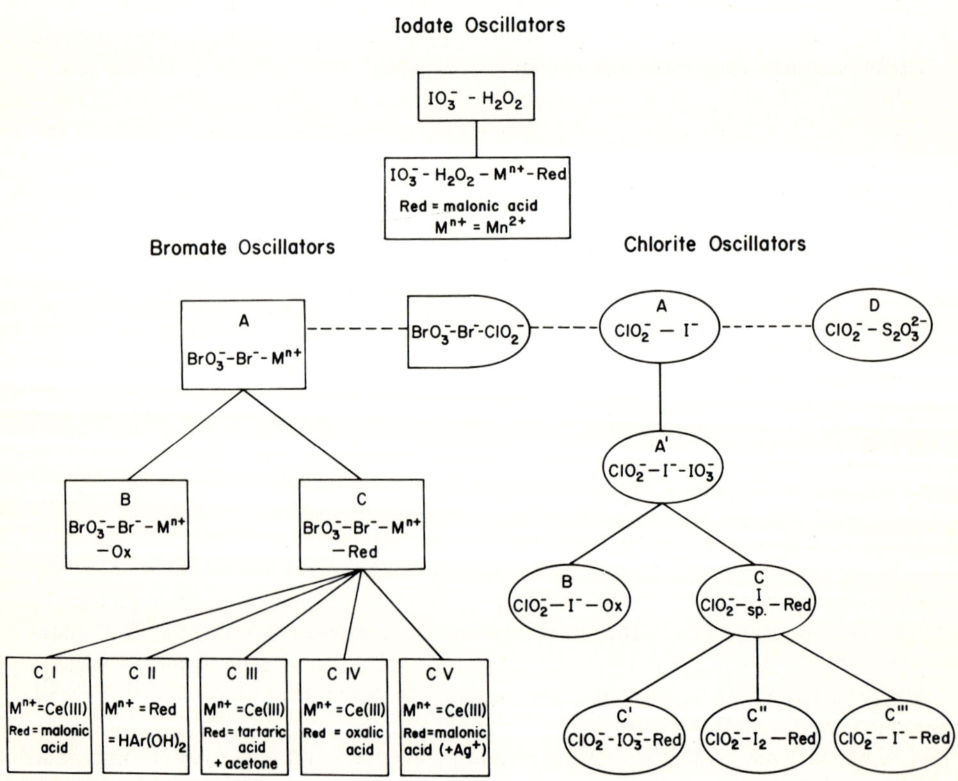

Fig. 17. Schematic classification of the known homogeneous oscillators

while the chlorite systems were developed and studied under flow conditions which greatly enhance the likelihood of observing oscillatory behavior.

Figure 17 contains a summary of the known homogeneous oscillators. The chlorite grouping was discussed in the previous section. The subcategories under C of the bromate section represent Noyes'[63] crucial insight in classifying the bromate oscillators known in 1980 according to mechanistic considerations. The recent discovery[65] of oscillations in a purely inorganic system containing only bromate, bromide and a redox couple enables us to place these organic bromate-driven oscillators in a more general context. It should soon prove possible to design purely inorganic bromate oscillators of Type C as well as of Type B, which have not yet been reported. The status of iodate oscillators is much less clear at this date, though the recent formulation[25,26] of a mechanism for the Briggs-Rauscher reaction as well as continuing work on the Bray-Liebhafsky system and the prospect of new oscillators of this type offer the hope of progress in the near future.

Future studies of chemical oscillators are likely to aim both at consolidating the gains already made and at expanding the range of phenomena under consideration. Investigation of the mechanism of chlorite oscillators, for example, is now well under way. A rapid kinetics study has been performed on the ClO_2^--I_2 reaction[53], leading to new insights, and further such studies particularly involving reactions of chlorine dioxide are likely to be necessary. Refining the crude characterization of chemical oscillators pictured in Fig. 17 will require the discovery of some of the "missing links" as well as further mechanistic developments. The discovery of oscillators based on non-halogen chemistry would broaden significantly the chemical scope of the field. Elements with multiple oxidation states, such as nitrogen, sulfur, carbon and perhaps some transition metals would seem to be logical starting points.

While this review has focused on homogeneous, isothermal, periodic reactions in the liquid phase, a variety of phenomena in which one or more of those restrictions is relaxed merit further study. For example, Gray and coworkers[66] have shown that the gas phase oxidation of a number of hydrocarbons in a flow reactor can exhibit several oscillatory modes as well as bistability, though temperature as well as concentration variations must be taken into account. Non-isothermal, or more properly, thermokinetic oscillations have also been reported in the liquid phase[67]. The oxidation of several gases at catalytic surfaces can proceed in an oscillatory fashion[68], and a similar phenomenon has been reported for the reaction of hydrogen with halogens and with oxyhalogen ions in acidic aqueous solution[69]. Designing new systems of these types and providing mechanisms for the existing ones provide many exciting challenges.

A number of reports of photochemically driven oscillators, in which a solution illuminated at constant intensity produces a periodically varying emission, have appeared[70–72]. However, recent work suggests that the source of oscillation in these systems is hydrodynamic rather than chemical, and that the photochemistry serves merely as a probe of the convective motion of the fluid[73]. A more interesting observation is that of Möckel[74], who has generated spatial structure by illuminating aqueous solutions containing potassium iodide, carbon tetrachloride and starch as well as various non-aqueous organic mixtures. The design of a genuine, photochemically driven homogeneous oscillator still remains an unsolved problem.

One phenomenon that is drawing increasing attention, not only from chemists, but from mathematicians and physicists as well, has been dubbed "chemical chaos", the existence of sustained aperiodic oscillation of species concentrations in a flow sys-

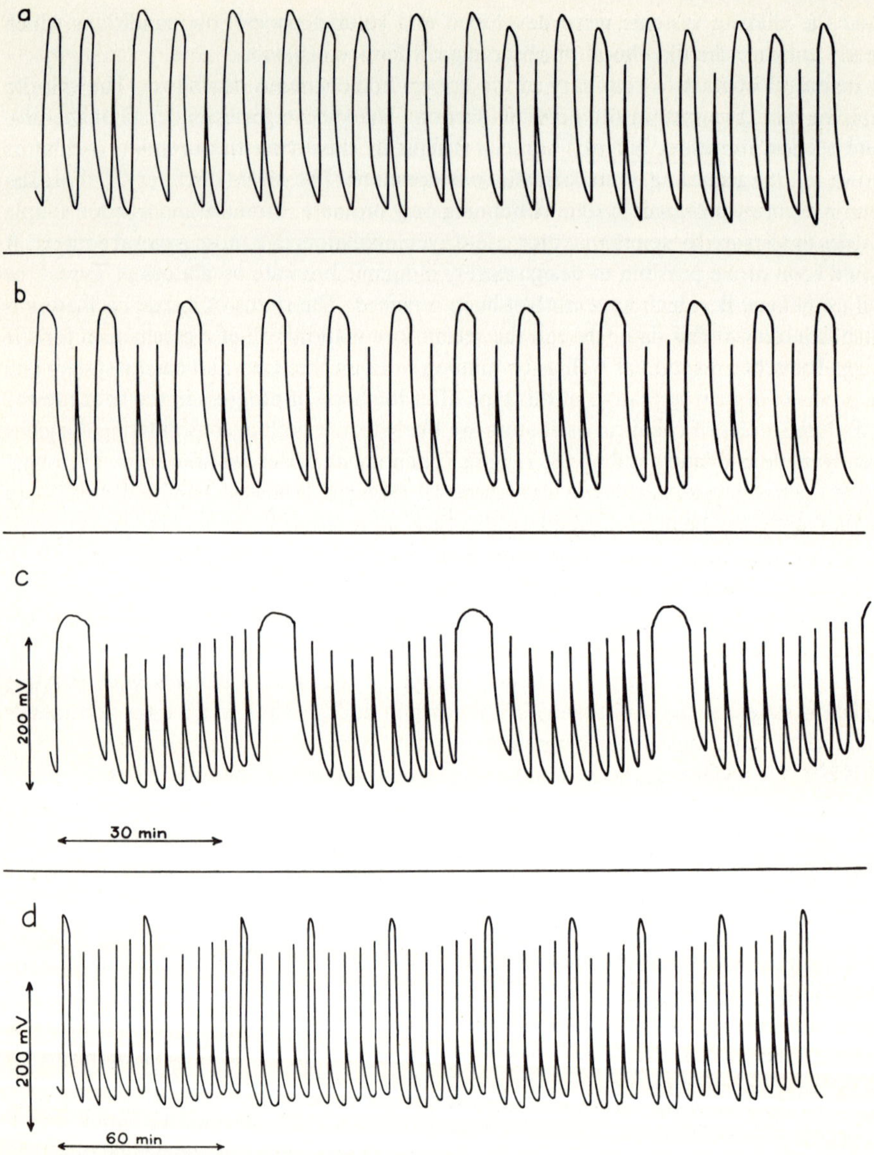

Fig. 18 a–d. Aperiodic oscillations in the chlorite-thiosulfate system in a CSTR. $[ClO_2^-]_0 = 5 \times 10^{-4}$ M, pH 4, T = 25.0 °C. **a–c** $[S_2O_3^{2-}]_0 = 3 \times 10^{-4}$ M, **d** $[S_2O_3^{2-}]_0 = 4 \times 10^{-4}$ M. Residence times: **a** 6.8 min, **b** 10.5 min, **c** 23.6 min, **d** 11.4 min

tem[75,76]. The origin of this behavior, illustrated in Fig. 18 for the chlorite-thiosulfate oscillator[77] is the subject of considerable controversy. One school of thought holds that it arises from random transitions, resulting from uncontrolled experimental fluctuations, between one mode of *periodic* oscillation and another[78]. The alternative view, supported by several model calculations[79], holds that chaotic behavior is *inherent in the dynamics* of certain complex chemical reaction networks and corresponds to the mathematical objects known as "strange attractors"[80]. A this stage, aperiodic behavior has been systematically studied in only two systems, the BZ[79, 81, 82] and the chlorite-thiosulfate[77] reactions, both of which give rise to some ambiguity. Further studies are therefore in order.

Finally, the quest to develop mechanistic explanations for these varied and fascinating phenomena can succeed only if more data become available on the component processes. Kinetics studies of the reactions which make up a complex oscillatory system are essential to its understanding. In some cases, traditional techniques may be adequate, though in many others, fast reaction methods will be required. There also appears to be some promise in developing an analysis of the relaxation of flow systems in non-equilibrium steady states as a technique to complement equilibrium relaxation techniques.

Acknowledgments. Work in this laboratory on oscillating chemical reactions has benefitted from the generous financial support of the National Science Foundation (Grant No. CHE 7905911) as well as from institutional Biomedical Research Support Grants, and a National Institute of Health Research Grant (GM 08893). Much of the work described was performed by several outstanding coworkers, most notably Patrick De Kepper, Miklós Orbán, Christopher Dateo, Mohamed Alamgir, Debra Banville, Robert Kaner and Steven Jacobs.

IX. References

1. Belousov, B. P.: Sb. Ref. Radiats. Med. 1958, Megdiz, Moscow, 145 (1959)
2. Morgan, J. S.: J. Chem. Soc., London, Trans. *109*, 274 (1916)
3. Bray, W. C.: J. Amer. Chem. Soc. *43*, 1262 (1921)
4. Liesegang, R. E.: Z. Phys. Chem. *52*, 185 (1905)
5. Winfree, A. T.: The Geometry of Biological Time. New York: Springer Verlag, 1980
6. McBirney, A., Noyes, R. M.: J. Petrology *20*, 487 (1979)
7. Smith, B. A. et al.: Science *215*, 504 (1982)
8. Gerisch, G., Hess, B.: Proc. Nat. Acad. Sci. U.S.A. *71*, 2118 (1974)
9. Gray, P. et al.: Proc. Roy. Soc. London *A 374*, 313 (1981)
10. Field, R. J., Noyes, R. M.: Acc. Chem. Res. *10*, 214 (1977)
11. Nicolis, G., Prigogine, I.: Self-Organzation in Nonequilibrium Systems. New York: John Wiley & Sons 1977
12. Neece, G. A., Widom, B.: Ann. Rev.P. Chem. *20*, 167 (1969)
13. Zhabotinsky, A. M.: Doklady Akad. Nauk SSR *157*, 392 (1964)
14. Ghosh, A., Chance, B.: Biochem. Biophys. Res. Commun. *16*, 174 (1964)
15. Gear, C. W.: Numerical Initial Value Problems in Ordinary Differential Equations. Englewood Cliffs, NJ: Prentice-Hall 1971
16. Hindmarsh, A. C.: Gear: Ordinary Differential Equation Solver. Tech. Report No. UCM-30001, Rev. 2. Lawrence Livermore Lab. 1972
17. Field, R. J. et al.: J. Amer. Chem. Soc. *94*, 8649 (1972)
18. Field, R. J., Noyes, R. M.: J. Chem. Phys. *60*, 1877 (1974)
19. Liebhafsky, H. A.: J. Amer. Chem. Soc. *53*, 896 (1931)
20. Briggs, T. C., Rauscher,W. C.: J. Chem. Educ. *50*, 496 (1973)
21. Orbán, M. et al.: J. Phys. Chem. *83*, 2056 (1979)
22. Edelson, D. et al.: Int. J. Chem. Kin. *7*, 417 (1975)

23. Sharma, K. R., Noyes, R. M.: J. Amer. Chem. Soc. *84*, 1355 (1962)
24. Odutola, J. A. et al.: J. Phys. Chem. *86*, 818 (1982)
25. De Kepper, P., Epstein, I. R.: J. Amer. Chem. Soc. *104*, 49 (1982)
26. Noyes, R. M., Furrow, S. D.: J. Amer. Chem. Soc. *104*, 45 (1982)
27. Onsager, L.: Phys. Rev. *37*, 405 (1931)
28. Lotka, A.: J. Amer. Chem. Soc. *42*, 1595 (1920)
29. Tyson, J. J.: J. Chem. Phys. *62*, 1010 (1975)
30. Schmitz, R. A.: Multiplicity, Stability, and Sensitivity of States in Chemically Reacting Systems – A Review. In: Adv. Chem., No. 148 (Hurlburt, H. M., ed.), Washington, D. C.: Amer. Chem. Soc. 1975
31. Franck, U. F.: Faraday Symp. Chem. Soc. *9*, 137 (1974)
32. Nitzan, A. et al.: J. Chem. Phys. *61*, 1056 (1974)
33. Boissonade, J., De Kepper, P.: J. Phys. Chem. *84*, 501 (1980)
34. Edelstein, B. B.: J. Theor. Biol. *29*, 57 (1970)
35. De Kepper, P. et al.: C. R. Acad. Sci. Ser. *C283*, 371 (1976)
36. De Kepper, P.: Ph. D. Thesis, Univers. Bordeaux, France 1978
37. Benson, S.: The Foundations of Chemical Kinetics. New York: McGraw-Hill Book Comp. 1960
38. Uppal, A. et al.: Chem. Eng. Science *29*, 967 (1974)
39. Jordan, P. C.: Chemical Kinetics and Transport. New York, Plenum Press 1979
40. De Kepper, P., Boissonade, J.: J. Chem. Phys. *75*, 189 (1981)
41. Landolt, H.: Ber. dtsch. chem. Ges. *19*, 1317 (1886)
42. Eggert, J., Scharnow, B.: Z. Elektrochem. *27*, 455 (1921)
43. Bognár, J., Sárosi, S.: Anal. Chim. Acta *29*, 406 (1963)
44. Roebuck, J. R.: J. Phys. Chem. *6*, 365 (1902)
45. Liebhafsky, H. A., Roe, G. M.: Int. J. Chem. Kinet. *11*, 693 (1979)
46. Papsin, G. A. et al.: J. Phys. Chem. *85*, 2575 (1981)
47. De Kepper, P. et al.: J. Amer. Chem. Soc. *103*, 6121 (1981)
48. Epik, P. A., Shub, N. S.: Dokl. Akad. Nauk SSSR *100*, 503 (1955)
49. Gribschaw, T. A. et al.: J. Phys. Chem. *85*, 2152 (1981)
50. Kern, D. M., Kim, C.-H.: J. Amer. Chem. Soc. *87*, 5309 (1965)
51. Meeus, J. de, Sigalla, J.: J. Chim. Phys. *63*, 453 (1966)
52. Epstein, I. R. et al.: J. Amer. Chem. Soc. *102*, 3751 (1980)
53. Grant, J. L. et al.: Inorg. Chem. *21*, 2192 (1982)
54. Dateo, C. et al.: J. Amer. Chem. Soc. *104*, 504 (1982)
55. De Kepper, P. et al.: J. Amer. Chem. Soc. *103*, 2133 (1981)
56. Försterling, H. D. et al.: Z. Naturforsch. *35a*, 1354 (1980)
57. Orbán, M. et al.: Nature *292*, 816 (1981)
58. Orbán, M. et al.: J. Amer. Chem. Soc. *104*, 5911 (1982)
59. De Kepper, P. et al.: J. Phys. Chem. *86*, 170 (1982)
60. Orbán, M. et al.: J. Phys. Chem. *86*, 431 (1982)
61. Winfree, A. T.: Sci. Amer. *230*(6), 82 (1974)
62. Marsden, J., McCracken, M: The Hopf Bifurcation and Its Applications. New York: Springer-Verlag 1976.
63. Noyes, R. M.: J. Amer. Chem. Soc. *102*, 4644 (1980)
64. Gordon, G. et al.: Prog. Inorg. Chem. *15*, 201 (1972)
65. Orbán, M. et al.: J. Amer. Chem. Soc. *104*, 2657 (1982)
66. Gray, P. et al.: Nonlinear Phenomena in Chemical Dynamics (Vidal, C., Pacault, A. eds.), Springer Ser. Synergetics, Vol. 12, New York: Springer-Verlag 1981, 20
67. Hafke, C., Gilles, E. D.: Mess. Steuern. Regeln. *11*, 204 (1968)
68. Sheintuch, M., Schmitz, R. A.: Catal. Rev. Sci. Eng. *15*, 109 (1977)
69. Orbán, M., Epstein, I. R.: J. Amer. Chem. Soc. *103*, 3723 (1981)
70. Nemzek, T. L., Guillet, J. E.: J. Amer. Chem. Soc. *98*, 1032 (1976)
71. Yamazaki, I. et al.: Photochem. Photobiol. *23*, 69 (1976)
72. Bose, R. J. et al.: J. Amer. Chem. Soc. *99*, 6119 (1977)
73. Valdes-Aguilera, O. et al.: to be published
74. Möckel, P.: Naturwissenschaften *64*, 224 (1979)
75. Ruelle, D.: Trans. N. Y. Acad. Sci. *35*, 66 (1971)

76. Rössler, O. E.: Z. Naturforsch. *31 a*, 259 (1976)
77. Orbán, M., Epstein, I. R.: J. Phys. Chem. *86*, 3907 (1982)
78. Ganapathisubramanian, N., Noyes, R. M.: J. Chem. Phys. *76*, 1770 (1982)
79. Turner, J. S. et al.: Phys. Lett. *85 A*, 9 (1981)
80. Hofstadter, D. R.: Sci. Amer. *245*(5), 22 (1981)
81. Hudson, J. L., Mankin, J. C.: J. Chem. Phys. *74*, 6171 (1981)
82. Vidal, C. et al.: Ann. N. Y. Acad. Sci. *357*, 377 (1980)

Synergic Extraction of Actinides

V. V. Ramakrishna and S. K. Patil

Radiochemistry Division, Bhabha Atomic Research Centre, Trombay, Bombay-400085, India

Studies on the solvent extraction of actinide ions by different combinations of extractants have been reviewed. Various equilibria involved in the extraction processes and the formation of the extractable complexes have been considered along with their equilibrium constant data. Various methods which are useful in establishing the composition and the nature of the extractable complexes are presented. The data on isolation and structural studies of some complexes, involved in synergic extraction, are also included. A brief description of the different areas in which synergic extraction is finding application is also given. Many combinations of extractants, where the studies conducted are very few but, which are likely to yield enhanced extractions are indicated. Areas of research, both from the academic and applied points of view, which require attention are suggested.

Structure and Bonding 56

A. Introduction

The word synergism, was first used by Coleman[1] to describe the phenomenon of the enhanced solvent extraction of metal ions by certain combinations of extractants. In an early study, it was stated[2] that synergism in the extraction seemed to be limited to dialkylphosphoric acid – neutral extractant combinations and, among the metal ions investigated, only to uranyl ion. The work reported much earlier[3], on the extraction of Pr(III) and Nd(III) by mixtures of a betadiketone(HTTA) and a neutral extractant (TBP) was overlooked. Very soon it was realized that synergic extraction is a general phenomenon with combinations of different types of extractants, as well as many metal ions, exhibiting it[4–11].

In solvent extraction the extractable metal complex is neutral, the charge on the metal ion being neutralized either by the anion from the aqueous medium employed for extraction or the anion provided by the extractant itself. If the neutral species is coordinatively unsaturated, it behaves as a Lewis acid and interacts with a viable donor (Lewis base) molecule to form a coordinatively saturated species. In the absence of any other donor water serves its role to form hydrated neutral species which are relatively poorly extractable into organic solvents. Either the charge neutralising ligand or the coordination saturating donor has to be aquophobic for the resulting metal species to become extractable into the organic solvents. In synergism the extracted complexes contain at least two types of ligands both of which are aquophobic thus enhancing extraction.

The opposite effect, viz., the combination of extractants giving a poor extraction was also noted simultaneously[2] and the word antagonism was suggested[12] to describe the phenomenon. While synergism, in general, is due to the formation of more extractable metal complexes with mixtures of extractants, antagonism is usually due to the interaction of the extractants with each other, resulting in a virtual removal of the available extractants from the system. With certain combinations of extractants, such as those observed in the extraction of U(VI) by mixtures of a carboxylic acid and a neutral donor[13], both these effects can take place depending on the experimental conditions used.

Though synergism is a part of general solvent extraction behaviour[14] it has retained its separate identity due to the complexity of the extraction equilibria, involving the formation of many complexes, and in view of the potential it has in promoting or suppressing extraction, as desired. Synergic extraction has attracted the attention of a large number of workers in view of its use in varied fields such as mutual separation and determination of metals, recovery of metals from complex matrices, determination of stability constants for metal ligand complexes in aqueous solutions, studies on the coordination capabilities of metal ions, preparation of compounds of interest in gas chromatography, partition chromatography etc. Though synergism has been investigated to some extent from fundamental, physicochemical and technological points of view, the areas unexplored are much more compared to what has been reported.

Since the presentation of the general reviews covering the earlier work[12, 15, 16] a review covering the subsequent work was presented in 1978[17]. Though the work on actinides alone was reviewed by Healy[18] the literature covered was only up to 1970. Some of the recent results reported on actinides have cleared many earlier doubts and as well opened up new routes of understanding. Hence a detailed discussion of the studies on synergic extraction systems involving actinides is considered useful for a better understanding of the metal complexing behaviour in general and metal extraction behaviour in particular. The literature that appeared till the end of 1982 was covered.

Besides presenting a general review of the literature it has been our objective to draw the attention of reseachers to this field which is both of practical and basic interest. Generally the compounds involved in synergic extraction are characterized employing the extraction methods alone. Thus, it is felt that there is a rich field here for structural chemists, especially those who can separate and purify the complexes, determine their structures and other physicochemical properties and then, through spectroscopic techniques, connect the isolated compounds with those in solution. Such studies appear promising as can be inferred from some recent work reported.

I. Measure of Synergism

Synergism has been expressed quantitatively by different ways. If D_A, D_B and $D_{(A+B)}$ respectively are the distribution ratios of a metal ion, under identical conditions, with extractants A, B and their mixture, containing the same concentration of each extractant used separately, it follows that

$$D_{(A+B)} = D_A + D_B \pm \Delta D \tag{1}$$

Thus, a positive value of ΔD shows synergism and a negative value of the same indicates antagonism, and its magnitude is a measure of the extent of these effects. The variation of ΔD, as a function of different experimental parameters, is the sole basis on which the composition of the extracted species, responsible for synergism, is usually obtained. The value of ΔD depends on many experimental parameters and, frequently ΔD approaches $D_{(A+B)}$, when $(D_A + D_B)$ is negligible relative to $D_{(A+B)}$.

The organic phase formation constants of adducts between metal chelate and the neutral donor, in the extraction with combinations of acid chelates and neutral extractants, have also been used as a measure of synergism[19]. When extended to other combinations of extractants such equilibria may not give a complete picture of synergism in view of many other competing equilibria.

The synergic coefficient, given as[20]

$$\text{S.C.} = \log \frac{D_{(A+B)}}{D_A + D_B} \tag{2}$$

is also a commonly used measure of synergism. Thus S.C. > 0 indicates synergism and S.C. < 0 suggests antagonism. The coefficient is given, sometimes, as

$$\text{S.C.} = \frac{D_{(A+B)}}{D_A + D_B} \tag{3}$$

The latter is used when the extent of synergism is only marginal. Both these measures are quite useful in choosing the experimental conditions for the extraction or back extraction of the metal ion.

II. Some Representative Equilibria

The equilibria usually encountered in the extraction of metal ions with combinations of betadiketones and neutral donors have been more widely investigated as compared with other systems and hence a detailed account of the same is given below.

The extraction of metal ions, from aqueous solutions, by betadiketones (HA) dissolved in organic solvents, is usually given as in Eq. (4) assuming that the neutral metal betadiketonate is the only extractable species, which, generally is the case.

$$M^{n+} + n\,HA_{(o)} \overset{K_A}{\rightleftharpoons} MA_{n(o)} + n\,H^+ \tag{4}$$

The subscript (o) represents the species present in the organic phase and all the other species are present in the aqueous phase. The equilibrium constant K_A, measured in concentration terms, is then given by

$$K_A = \frac{[MA_n][H^+]^n}{[M^{n+}][HA]^n} \tag{5}$$

where the square brackets denote concentrations. If C_M is the total concentration of the metal ion in the aqueous phase, at equilibrium, it can be shown that

$$C_M = [M^{n+}]F \tag{6}$$

where

$$F = 1 + \sum_1^j \beta_{OH(j)}[OH^-]^j + \sum_1^j \beta_{A_{(j)}}[A^-]^j + \sum_1^j \beta_{L_{(j)}}[L^-]^j \tag{7}$$

and $\beta_{OH(j)}$, $\beta_{A_{(j)}}$ and $\beta_{L_{(j)}}$ are the stability constants for the equilibria represented by Eqs. (8), (9) and (10), respectively,

$$M^{n+} + j\,OH^- \rightleftharpoons M(OH)_j^{(n-j)+} \tag{8}$$

$$M^{n+} + j\,A^- \rightleftharpoons M(A)_j^{(n-j)+} \tag{9}$$

$$M^{n+} + j\,L^- \rightleftharpoons M(L)_j^{(n-j)+} \tag{10}$$

and OH^- and L^- are the hydroxide ion and the anion present in the aqueous medium respectively. The formation of mixed ligand metal complexes of the type $M(OH)_jA_kL_l$ is ignored for the sake of simplicity. Thus F is the factor that accounts for the presence of metal ion in aqueous phase in the form of its different complexes. Incorporating Eq. (6) in Eq. (5), we get

$$K_A = \frac{[MA_n][H^+]^nF}{C_M[HA]^n} = \frac{D_A[H^+]^nF}{[HA]^n} \tag{11}$$

where D_A is the observed distribution ratio of the metal ion between the two phases, when HA is used as the extractant.

The extraction of metal ions by a neutral extractant (B), such as neutral organophosphorus compounds dissolved in organic solvents, is usually given as in Eq. (12),

$$M^{n+} + n\,L^{-} + m\,B_{(0)} \overset{K_B}{\rightleftharpoons} ML_n \cdot m\,B_{(0)} \tag{12}$$

where L^- is the anion of the aqueous medium (only monocharged anion is considered for the sake of simplicity) from which the extraction is carried out. The equilibrium constant K_B is then given by

$$K_B = \frac{D_B \cdot F}{[L^-]^n[B]^m} \tag{13}$$

where D_B is the observed distribution ratio of the metal ion when B is used as the extractant and F has the same meaning as explained before.

When combinations of HA and B are employed for the metal ion extraction, in addition to those given in Eqs. (4) and (12), the following equilibria need be considered,

$$M^{n+} + (n-1)L^{-} + HA_{(o)} + a\,B_{(o)} \rightleftharpoons ML_{(n-1)}A \cdot a\,B_{(o)} + H^{+} \tag{14}$$

$$M^{n+} + (n-2)L^{-} + 2\,HA_{(o)} + b\,B_{(o)} \rightleftharpoons ML_{(n-2)}A_2 \cdot b\,B_{(o)} + 2\,H^{+} \tag{15}$$

etc., and

$$M^{n+} + n\,HA_{(o)} + y\,B_{(o)} \overset{K_{n,y}}{\rightleftharpoons} MA_n \cdot y\,B_{(o)} + n\,H^{+} \tag{16}$$

The extracting species shown in Eq. (16) is the adduct of the metalbetadiketonate with the neutral donor molecules. It is now generally accepted that such adducts are formed in the organic phase by equilibria such as,

$$MA_{n_{(0)}} + y\,B_{(0)} \overset{\beta^0_{n,y}}{\rightleftharpoons} MA_n \cdot y\,B_{(0)} \tag{17}$$

It is seen that Eq. (17) results by substracting Eq. (4) from Eq. (16), thus showing that such organic phase adduct formation equilibria can be derived from extraction data, using the relationship,

$$\frac{K_{n,y}}{K_A} = \beta^0_{n,y} \tag{18}$$

The extracting species such as those shown in Eqs. (14) and (15) are mixed quaternary adducts and they can be visualised as forming from the ternary adduct, $MA_n \cdot yB$, by replacing the betadiketonate anion by the anion present in the aqueous medium. Alternatively they can be formed by replacing the anion from the adduct $ML_n \cdot mB$ by the betadiketonate anion. It was suggested[3, 21)] that complexes can also be formed in nonaqueous solvents by stepwise replacement of one ligand by another.

Another mode of extraction that is very likely in these systems is the extraction of ion-pairs such as $[MA_{(n-1)} \cdot aB]^+[L]^-$. These adducts are different from those described in

Eqs. (14) and (15), in the sense that the anion L^- is outside the coordination sphere of the metal ion in the ion-pairs.

III. Experimental Methods

A number of methods are used for establishing the composition of the extracted metal species and for the calculation of the relevant equilibrium constants. The application of these methods is illustrated below using the HA-B extractant combinations. A similar approach can be used for other systems as well with suitable modifications dictated by any additional equilibria that may be involved in them.

The use of metal ions at trancer concentrations, if possible by employing radioactive tracers, enables one to carry out such extractions where $C_M \ll [HA]$ or [B], thereby permitting the approximation, that the equilibrium concentrations of either extractants are the same as their initial concentrations. Besides, use of radioactive tracers makes the distribution measurements convenient and fast. Such a condition makes the quantitative interpretation of the extraction data much simpler when the slope-analysis, Job's and the curve-fitting methods, described below, are employed.

Most of the studies involving HA-B combinations were conducted employing such aqueous media, the extraction from which involves the formation of simple ternary adducts $MA_n \cdot yB$ and their study can easily be carried out with any of the methods given here. But when quarternary adducts or ion-pairs are involved in the extraction it becomes very difficult to establish their stoichiometry by any single method and very often different types of experimental data need be obtained to confirm such species.

(a) Slope-Analysis Method

The distribution ratio $D_{(A+B)}$ is obtained as a function of one of the variables, $[H^+]$, [HA] or [B], keeping the other two constant at a time. The composition of the adduct can then be established from the slope analysis[1] of the plot of log ΔD vs. log[variable]. The data can be used for the calculation of $K_{n,y}$, the equilibrium constant represented by Eq. (16) and $\beta^0_{n,y}$, that represented by Eq. (17). This method is widely applied in extraction studies involving all types of combinations of extractants. The method is generally applicable when only one synergic species is predominant under the experimental conditions used. The limitations of this method arise in interpreting the data when non-integral slope values are obtained, which may be due to different reasons such as the interaction of the extractants with each other or due to the extraction of different species of varying composition.

(b) Job's Method

Gal and Nikolic[22] have shown that Job's method of continuous variations could be successfully used for finding out the composition of the species responsible for synergism and also for calculating the equilibrium quotients. In this method the distribution ratio data are obtained using a fixed composition of the aqueous phase and employing mixtures of HA and B in different ratios, but, keeping their total concentration constant. The composition of the extracted species can then be arrived at from the ratio of [HA] : [B] for which the ΔD value reaches the maximum. The value of $K_{n,y}$ can also be calculated from these data. This method is applicable to systems where only one species is responsible for synergic enhancement in the extraction of the metal ion.

(c) Curve-Fitting Methods

When a series of adducts $MA_n \cdot B$, $MA_n \cdot 2B$ etc. are responsible for the observed synergism their composition as well as their formation constants can be determined by obtaining the metal ion distribution ratio data using a fixed [HA] and as a function of increasing [B]. From Eqs. (4), (16) and (17) it can be shown[23] that

$$\frac{\Delta D}{D_A[B]} = \beta^0_{n,1} + \beta^0_{n,2}[B] + \text{etc.} \tag{19}$$

This method has been widely used where simple ternary adducts are the species responsible for synergism.

In some systems a series of adducts $ML_{(n-1)}A \cdot aB$, $ML_{(n-2)}A_2 \cdot bB, \ldots, MA_n \cdot yB$ are involved in the extraction. An attempt was recently made[24, 25] to determine their composition and the corresponding equilibrium constants by obtaining the metal ion distribution ratio data using a fixed [B] and varying [HA].

(d) Mole Ratio Method

The extraction of a metal ion is studied as a function of one of the variables [HA] or [B], keeping the concentration of the other in excess as compared to that of the metal ion. The extraction of the metal ion is then followed as a function of [HA]/[M] or [B]/[M] and the ratio for the maximum extraction gives the stiochiometry of the extracted adduct[26, 27]. Choice of proper experimental conditions and a high value for the adduct formation constant are necessary for successful application of this method.

(e) Organic Phase Adduct Formation Studies

These methods involve the study of the organic phase reactions such as the one represented by Eq. (17). If the metal chelate, neutral donor and their adduct show difference in their physicochemical properties, the same can be exploited for finding out the adduct composition and obtaining the corresponding equilibrium constants. Such studies have been reported which utilise the measurement of absorption spectra[28–30], changes in the epr parameters of the metal ion[31], induced nmr shift in the spectra of neutral donor due to the metal chelate[32–34], infrared measurements[35], changes in the solubility in an inert diluent[36], measurement of heat of reaction[37, 38] etc. As the data are independent of the solvent extraction behaviour, these are highly useful for corroborating the conclusions derived from the extraction data.

IV. Interactions

When combination of different extractants, together with a diluent, are employed for the metal ion extraction it is essential to know various intermolecular interactions that take place between them in order to successfully apply the law of mass action to these systems. These include (i) self-association, (ii) association with the diluent and (iii) association between different extractants. While the first two influence the extraction behaviour involving even a single extractant, all of them influence synergism markedly. Some of the representative data are discussed below which enable one to understand the extent of such interactions.

(a) Extractants and Diluents

Diluents are normally used to keep the extractants in solution and for suitably adjusting the extractant characteristics such as concentration, density, viscosity etc. Normally the diluent alone is a very poor extractant, if at all, and is expected to be an inert participant in the extraction of the metal ion. But diluents seldom behave as expected and they alter the extraction characteristics of the extractant appreciably.

A detailed information regarding the distribution of each of the extractants between a given aqueous medium and an organic diluent is essential, not only for knowing the changes from initial to equilibrium values of extractant concentration, but also for understanding the extractant-diluent interaction.

The partition data for the dilute solution of the extractants TBP, TBPO, TOPO, TEHPO, TPPO and some betadiketones were reported[39–44] and some of these data are given in Table 1. The distribution of TPrP between water and several diluents was studied[45] and the data were used for calculation of the equilibrium constant for the reaction represented by

$$TPrP_{(o)} + S_{(o)} \overset{K}{\rightleftharpoons} TPrP \cdot S_{(o)} \tag{19}$$

where S represents the diluent molecule, and the values of K obtained are given in Table 2. The association of TBP with different diluents was studied using infrared measurements[46, 47] and the values of K obtained[47] are also included in Table 2. Thus, these data clearly demonstrate that diluents used are seldom "inert" and that they interact with the extractants to different extent and this has pronounced influence on the activities of the donors. In general, equilibria involving extractants are expressed using concentrations rather than the activity, and are quoted for a diluent. However, some attempts were made to get "diluent free" equilibrium values by taking the above interactions into consideration[48, 49].

(b) Different Extractants

The interactions between different extractants are usually inferred either from the changes in the distribution ratio of one of the extractant, kept at fixed concentration,

Table 1. Partition data of some extractants[40, 43] aqueous phase – 0.10 M $(Na,H)ClO_4$; temp –25 °C

Diluent	log P_{HTTA}	log P_{TBP}	log P_{TBPO}
n-hexane	0.68	2.73	−0.52
n-heptane	0.57	2.75	−0.60
Cyclohexane	0.56	2.90	−0.22
Methylene chloride	1.84	4.50	2.48
Chloroform	1.73	4.20	3.08
CCl_4	1.30	3.70	0.77
Benzene	1.62	3.93	0.92
Toluene	1.60	3.84	0.69
Isopropylbenzene	1.37	3.50	0.24
Chlorobenzene	1.65	4.04	1.28
O-dichlorobenzene	1.59	4.07	1.34
Bromoform	–	4.56	2.32

Table 2. Association constants of TPrP/TBP with different diluents

Diluent	K	
(S)	TPrP[45]	TBP[47]
Benzene	–	0.2
MIBK	1.0	0.8
CCl_4	0.8	0.9
TBP	2.4	2.9
$CHCl_3$	5.4	6.1

with increasing concentration of the other extractant, or from the changes in the absorption spectrum of one of the extractant by the addition of other extractant.

Betadiketones (HA) were shown[50–53] to interact with neutral donors (B) forming 1 : 1 adducts, as represented by

$$HA_{(o)} + B_{(o)} \xrightleftharpoons{K} HA \cdot B_{(o)} \tag{20}$$

and some values of the equilibrium constants reported are given in Table 3. Evidently, for a given acid, higher the basicity of the donor more stable is the adduct.

Betadiketones were shown to interact with amines[54–57] and the data obtained[57] on the interaction of HTTA with QX, a quarternary amine, for the equilibrium represented by

$$QX_{(o)} + HTTA_{(o)} \xrightleftharpoons{K_X} Q(TTA)_{(0)} + H^+ + X^- \tag{21}$$

are given in Table 4. It is seen that the anion associated with the amine salt plays a dominating role in deciding the extent of this interaction.

HBPHA, a chelating agent and the donors TBP and TOPO were found[58] to associate forming 1 : 1 adducts and the adduct formation constants, determined in several diluents,

Table 3. Equilibrium constant data for 1 : 1 complexes between betadiketones and neutral donors

Betadiketone	Neutral donor	Medium	log K	Ref.
HHFA	TBP	CCl_4	0.26	50
HTFA	TBP	CCl_4	−0.65	50
HTTA	TBP	CCl_4	−0.72	50
HHFA	DEEP	CCl_4	0.44	50
HTFA	DEEP	CCl_4	−0.52	50
HTTA	DEEP	CCl_4	−0.57	50
HTTA	TBP	Kerosene	0.54	51
HTTA	TBP	C_6H_6	0.46	52
HTTA	TBP	Xylene	0.41	53
HTTA	DOSO	Xylene	0.56	53
HTTA	DBBP	Xylene	0.94	53
HTTA	TOPO	Xylene	1.24	53

Table 4. Equilibrium data for the interaction of HTTA with QX[57]

X	$-\log K_x$
Cl^-	1.30
Br^-	2.55
NO_3^-	3.24
I^-	4.50
SCN^-	5.44
ClO_4^-	6.46

Table 5. Association constants for the adducts of HBPHA and TBP or TOPO[58]

Diluent	P_{HBPHA}	$\log K_{TBP}$	$\log K_{TOPO}$
CCl_4	14.4	1.24	2.41
1,2-dichloroethane	128	0.71	1.83
1,1,2,2-tetrachloroethane	292	0.27	1.20
1,1,1-trichloroethane	14.2	1.24	–
Nitrobenzene	98.9	0.90	–
Chlorobenzene	50.0	1.06	2.22
m-xylene	19.3	1.27	2.51
Toluene	29.3	1.17	2.43
n-hexane	0.93	2.03	3.34
Benzene	40.5	1.11	2.32
Diisobutylketone	56	0.73	1.83
MIBK	178	0.13	1.25
Diethylketone	219	0.05	–
Dibutylether	13.4	1.12	2.28
Diisopropylether	25.9	0.93	2.09
n-butylacetate	122	0.41	–
n-propylacetate	144	0.22	–
Chloroform	203	–	–

are given in Table 5. The association constants are seen to decrease in diluents where the distribution constants of HBPHA are high. This behaviour is expected as the diluent also interacts with HBPHA, thus competing with the neutral donor.

Carboxylic acids interact extensively with neutral donors and in a typical study[59] the equilibrium constant for the association of benzoic acid (HB_Z) with TOPO given by the equilibrium

$$HB_{Z(o)} + TOPO_{(o)} \rightleftharpoons HB_Z \cdot TOPO_{(o)} \tag{22}$$

was found to be 6.1×10^3 in CCl_4. Carboxylic acids were also found[60] to interact with amine salt extractants.

Neutral donors and amine salts were shown[61] to interact with each other using an infrared method.

The interactions of some neutral extractants with some organophosphorus acids have been extensively studied and two excellent reviews[62, 63] have summarised the results.

Thus, there is hardly any extractant combination which is free from mutual association and this factor has to be taken into account in any quantitative study on synergic extraction.

B. Chelating Acid – Neutral Donor Mixtures

I. Oxycations

(a) Uranium(VI)

The hexavalent state of uranium ion is the usually encountered ion in the solution chemistry of uranium and its exceptional stability, relative to its other oxidation states as well as to other hexavalent actinide ions, makes studies with this ion much simple.

Though chelating acids, in general, extract U(VI) by the equilibrium,

$$UO_2^{2+} + 2\,HA_{(o)} \overset{K_A}{\rightleftharpoons} UO_2A_{2(o)} + 2\,H^+ \tag{23}$$

an additional equilibrium,

$$UO_2^{2+} + 3\,HA_{(o)} \overset{K_{AB}}{\rightleftharpoons} UO_2A_2 \cdot HA_{(o)} + 2\,H^+ \tag{24}$$

is also involved, in some cases, depending on the experimental conditions and the chelating acid used. An undissociated chelating acid HA is seen to behave as a neutral donor and form an adduct with the metal chelate, as given by

$$UO_2A_{2(o)} + HA_{(o)} \overset{\beta^0_{AB}}{\rightleftharpoons} UO_2A_2 \cdot HA_{(0)} \tag{25}$$

Table 6. Equilibrium constant data for the chelating acid adducts of uranyl chelates

Chelating acid	Aqueous medium	Diluent	$-\log K_A$	$-\log K_{AB}$	$\log \beta^0_{AB}$	Ref.
HAA		Chloroform	6.70	5.54	1.16	64
HAA	$NaClO_4$ (I = 1)[a]	Benzene		4.69		66
HBA		CCl_4		5.06		67
HDBM		Benzene		4.12		68
HTTA		Benzene	2.40	1.90	0.50	72, 74
HIPT	0.1 M $NaClO_4$	Chloroform	2.63	2.63		75
HIPT	$NaClO_4$	Benzene		3.49		76
HO_x	0.1 M $KClO_4$	Chloroform		1.60		77
	0.1 M $NaClO_4$	1,2-dichloro-ethane		0.70		80

[a] Values at different ionic strengths are given[66)]

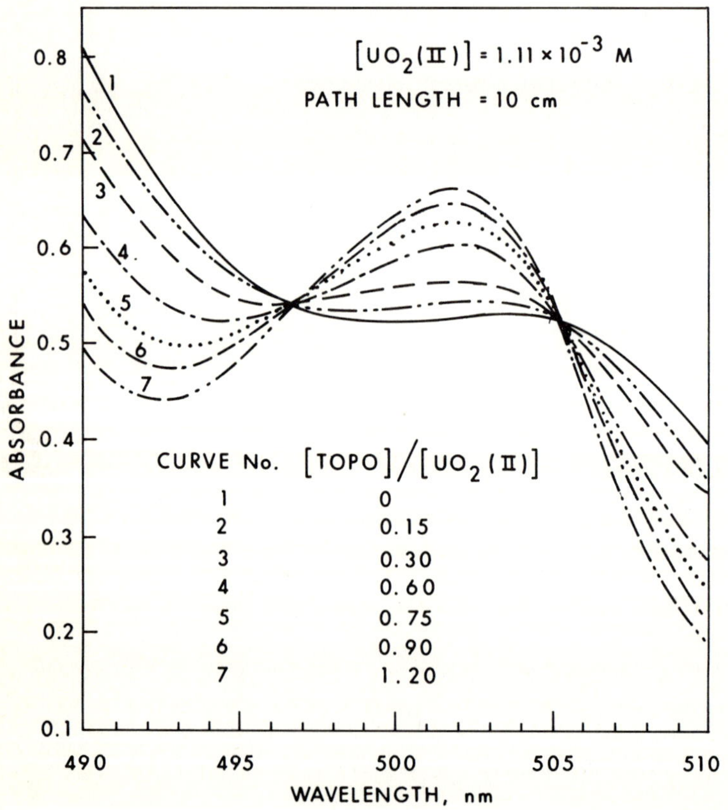

Fig. 1. Absorption spectra of U(VI) in excess HTTA as a function of different amounts of TOPO. (Reprinted from Ref. 96 by courtesy of Indian Academy of Sciences, Bangalore)

This behaviour is similar to that observed in the extraction by chelating acid – neutral donor combinations, where such adducts are extracted. This is exhibited by U(VI) in the extraction with the reagents such as HAA[64–66], HBA[67], HDBM[68, 69], HTTA[70–74], HIPT[75, 76] and HO_x[77–80]. A summary of the equilibrium constants, for these systems, are given in Table 6. It appears that those chelating acids which form more stable metal chelates may also behave as strong neutral donors.

The extraction of U(VI) by HTTA-B mixtures was studied by a large number of workers[5, 9–11, 81–85]. All these studies agree that the extraction of a 1:1 adduct, $UO_2(TTA)_2 \cdot B$, is responsible for the synergism, according to the equilibrium,

$$UO_2^{2+} + 2\,HTTA_{(0)} + B_{(0)} \overset{K_{2,1}}{\rightleftharpoons} UO_2(TTA)_2 \cdot B_{(0)} + 2\,H^+ \tag{26}$$

Though the extraction of $UO_2(TTA)_2 \cdot 2$ B and even $UO_2(TTA)_2 \cdot 3$ B were suggested in the earlier studies[5, 9, 10, 213, 214], subsequent results[86, 87, 96] have firmly established the absence of the formation of any such adducts. The absorption spectrum of $UO_2(TTA)_2$ in benzene and the changes caused to it by the addition of different amounts of TOPO are shown in Fig. 1. The isosbestic points observed suggest the formation of only one adduct and its composition as $UO_2(TTA)_2 \cdot B$ is clearly seen from the absorption data shown in Fig. 2.

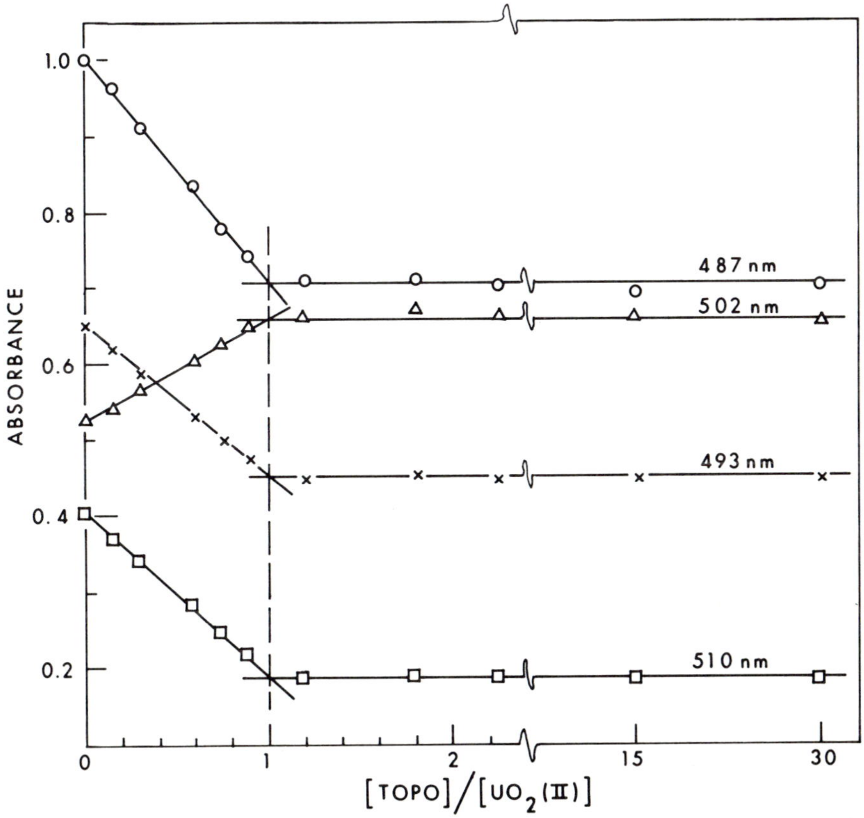

Fig. 2. Variation of the absorbance of U(VI) in HTTA as a function of [TOPO]/[U(VI)]. (Reprinted from Ref. 96 by courtesy of Indian Academy of Sciences, Bangalore)

Table 7. Equilibrium constant data for the extraction of U(VI) by mixtures of HTTA and neutral donors

Neutral donor	Diluent	Aqueous medium	$-\log K_A$	$\log K_{2,1}$	$\log \beta^0_{2,1}$	Ref.
TBPO	Cyclohexane	0.01 M HNO_3	2.8	3.9	6.7	5
TBP	Cyclohexane	0.01 M HNO_3	2.8	3.56	6.36	5
TPP	Benzene	0.01 M HCl	–	−0.32	–	9
TBP	Benzene	0.01 M HCl	–	2.48	–	9
TPPO	Benzene	0.01 M HCl	–	3.54	–	9
TBP	Benzene	2.0 M $LiClO_4$	2.40	2.65	5.05	81
TBP	Benzene	0.1 M $NaClO_4$	2.68	2.86	5.54	83
TBPO	Benzene	0.1 M $NaClO_4$	2.68	5.82	8.50	86
TOPO	Benzene	0.1 M $NaClO_4$	2.68	5.78	8.46	87
2,6-DMPO	Chloroform	dil. HCl (pH = 1.4)	2.77	−0.14	2.63	93
2,6-DMPO	Chloroform	dil. HCl (pH = 1.4)	–	–	2.60	92
PO	Chloroform	dil. HCl (pH = 1.4)	–	–	2.86	92
2 MPO	Chloroform	dil. HCl (pH = 1.4)	–	–	3.05	92
3 MPO	Chloroform	dil. HCl (pH = 1.4)	–	–	3.24	92
4 MPO	Chloroform	dil. HCl (pH = 1.4)	–	–	3.52	92
TLAO	Benzene	HNO_3	2.83	5.04	7.88	94
TBP	Benzene	1 M $NaClO_4$	1.86	2.78	4.64	95
TBP	Benzene	1 M HNO_3	–	2.88	–	95
TBP[a]	Benzene	1 M $HClO_4$	1.51	2.92	4.43	98
TBP[b]	Benzene	1 M $NaClO_4$	1.67	2.46	4.13	98
TBP[b]	Cyclohexane	0.01 M HNO_3	1.54	3.13	4.67	6

[a] – Np(VI); [b] – Pu(VI)

Table 8. Equilibrium constant data for the extraction of U(VI) by mixtures of HTTA and neutral donors[40, 83, 86, 87)]

Diluent	$-\log K_A$	TBP		TBPO		TOPO	
		$\log K_{2,1}$	$\log \beta^0_{2,1}$	$\log K_{2,1}$	$\log \beta^0_{2,1}$	$\log K_{2,1}$	$\log \beta^0_{2,1}$
n-hexane	3.18	3.49	6.67	5.78	8.96	6.40	9.58
n-heptane	2.91	3.56	6.47	5.72	8.63	6.40	9.31
Cyclohexane	2.86	3.96	6.81	6.32	9.18	6.66	9.52
Methylene-chloride	2.50	1.92	4.42	4.50	7.00	4.53	7.03
Chloroform	2.93	1.53	4.46	4.09	7.02	3.86	6.79
CCl_4	3.04	3.02	6.06	5.60	8.64	5.81	8.85
Benzene	2.68	2.96	5.54	5.82	8.50	5.78	8.46
Toluene	2.85	2.85	5.70	5.87	8.72	5.77	8.62
Isopropyl-benzene	2.85	2.54	5.39	5.56	8.41	5.56	8.41
Chlorobenzene	2.71	2.72	5.43	5.75	8.46	5.77	8.48
O-dichlorobenzene	2.70	2.92	5.62	5.82	8.52	5.80	8.50
Bromoform	2.80	2.06	4.86	4.38	7.18	4.27	7.07

Table 9. Equilibrium constant data for the extraction of U(VI) by HA-B mixtures[81, 82, 93)]

No.	HA	$-\log K_A$	$\log K_{2,1}$	$\log \beta^0_{2,1}$
1.	HFTA	2.25	2.85	5.10
2.	HTTA	2.40	2.65	5.05
3.	HBTA	2.95	2.40	5.35
4.	HTA	4.70	−1.35	3.35
5.	HAA	5.85	−2.25	3.60
6.	HBA	5.50	−1.65	3.85
7.	HDBM	5.50	−1.05	4.45
8.	HTTMA	–	−2.66	–
9.	HDBM	4.7	−1.9	2.8
10.	HDBM	4.7	−4.1	0.6
11.	HHFA	1.41	1.79	3.20
12.	HBTA	2.71	−0.15	2.56
13.	HTTA	2.77	−0.14	2.63
14.	HTFA	3.19	−0.71	2.48

Sr. No. 1–8: diluent-benzene, aqueous medium – 2.0 M $LiClO_4$ – B – TBP
Sr. No. 9–10: diluent-benzene, aqueous medium – B – TBP; for 10, B – EHA
Sr. No. 11–14: diluent-chloroform, aqueous medium – dil. HCl (pH = 1.4) – B – 2,6 – DMPO

In an early study of these systems[5)], employing nitric acid as the aqueous medium, the species $UO_2(TTA)(NO_3) \cdot B$ was also reported to be involved in the extraction. But a recent study[95)] using the same medium for extraction conclusively established the absence of the extraction of any such species and this observation was further supported by a spectrophotometric study[96)].

The various equilibrium constants for the equilibria involved in the extraction of U(VI) by HTTA-B combinations are summarized in Table 7. A comparison of the data is rather difficult in view of the different experimental conditions employed. The deviation of 2,6-DMPO and 2 MPO from the basicity order of the series was attributed[92)] to the occurence of steric effects with these donors.

The effect of diluents on the extraction of $UO_2(TTA)_2 \cdot B$ was studied[10, 83–87)] and the adduct composition was reported to remain unaltered with change in diluent. However, the diluents have marked influence on the equilibrium constants as can be seen from the data given in Table 8. Attempts were made[83)] to determine diluent-free equilibrium constants by determining the activities of the solutes in different diluents.

In an attempt to resolve the suggestion[19)] that in the adduct $UO_2(A)_2 \cdot B$ the B is not coordinated to the metal ion but to some side position, the adduct formation data were obtained[81)] with various structurally differing betadiketones and these are given in Table 9. As the extraction of the adducts was seen to run parallel to the extraction of the metal chelates and not influenced by the chelating acid structure, it was concluded[81)] that the neutral donor was directly attached to the metal ion. More equilibrium constant data[82, 93)] are included in Table 9. The data in this table do not clearly show the expected trend, viz., stronger the metal chelate, weaker is its adducts[12)], thus requiring further confirmation.

A mixed species, $UO_2(AA)(ClO_4) \cdot B$, was suggested[97)] to be involved in the extraction of U(VI) from perchlorate medium by HAA-B mixtures. Further study is required

Table 10. Equilibrium constant data[76] for the equilibrium (Eq. 27)

Neutral donor	$\log \beta^0$
TBP	0.30
DOSO	0.75
DBBP	1.27
TPPO	1.48
TOPO	3.13

to confirm this especially in view of the absence of any such mixed species in the extraction from nitrate medium[95, 96].

Moderately strong neutral donors were reported[76] to replace the additional chelating acid molecule and give synergism in the extraction of U(VI) by HIPT-B mixtures, according to the equilibrium,

$$UO_2(IPT)_2 \cdot HIPT_{(o)} + B_{(o)} \overset{\beta^0}{\rightleftharpoons} UO_2(IPT)_2 \cdot B_{(o)} + HIPT_{(o)} \tag{27}$$

and the reported values of the equilibrium constants are given in Table 10.

(b) Neptunium(VI) and Plutonium(VI)

The hexavalent state of neptunium and plutonium are identical to that of U(VI), in the sense that both exist as MO_2^{2+} in aqueous solutions. Whereas U(VI) is quite stable in aqueous solutions Np(VI) and Pu(VI) are unstable, especially when in contact with organic solutions, and need to be stabilized by using holding oxidants. Only a few studies have been reported for these systems. In the extraction with HTTA-B mixtures, the species $MO_2(TTA)_2 \cdot B$ was reported[6, 98] to be responsible for the observed synergism with both these ions, just like U(VI), and the equilibrium constant data are included in Table 7. It appears that the stabilities of the $MO_2(TTA)_2 \cdot B$ adducts of U(VI), Np(VI) and Pu(VI) are of, more or less, similar magnitude.

In the extraction of Pu(VI) by TBP in presence of various betadiketones[99] the order of extraction followed HTTA > HBTA > HTFA > HHFA which is the same as that observed for U(VI).

(c) Neptunium(V)

Though Np(V) is exceptionally stable in aqueous solutions, as compared to its neighbours U(V) and Pu(V), there was little attempt to take advantage of it, for extraction studies. However, the only study[6] on its extraction with HTTA-TBP was plagued with the problem of getting reproducible data. Though synergism was observed, the species suggested, viz., $HNpO_2(TTA)_2 \cdot TBP$ is more speculative and the equilibrium constants reported are uncertain.

II. Tetravalent Ions

It was predicted[5)] that there would be no synergic effect with tetravalent metal ions, in the extraction with HA-B mixtures, as it was felt that the chelates of these metal ions with bidentate ligands are coordinatively saturated with a coordination number of eight. However, synergism was soon observed for these systems as well.

Pronounced tendency of tetravalent actinides, especially of Np(IV) and Pu(IV) to undergo hydrolysis and polymerization and their instability of oxidation state in low acidic media, make it necessary to carry out their extraction studies from fairly high aqueous acidic media. Depending on the aqueous media and the neutral donor used, several species, involving the anion present in the aqueous media, can participate in the extraction. This, therefore, calls for a careful interpretation of extraction data.

(a) Thorium

The extraction of thorium into HTTA-B mixtures was conducted by many workers[7, 9–11, 48, 94, 100–105)] and except those using nitric acid as aqueous medium for extraction all others agree that the adduct $Th(TTA)_4 \cdot B$ is mainly responsible for synergism. Though adducts of higher order such as $Th(TTA)_4 \cdot 2B$ were also suspected[48)], the subsequent studies did not provide any support for the same. The higher than unity slope values, for the log D − log [B] plots, observed in chloroform diluent[10, 48)] might be due to the extraction of an additional species such as $Th(TTA)_3(B)_j^+L^-$, as such ion-pairs were reported[106, 107)] in the extraction of lanthanides.

In the studies carried out using nitric acid as the aqueous medium, the nitrate bearing species $Th(TTA)_3(NO_3) \cdot TBP$[7, 100)], $Th(NO_3)_2(TTA)_2 \cdot 2\,TBP$[7)], $Th(NO_3)_3(TTA) \cdot TBPO$[7)], $Th(NO_3)(TTA)_3 \cdot TLAO$[94)], $Th(NO_3)_2(TTA)_2 \cdot 2\,MNDDA$[101)] were suggested to be responsible for synergism. Though doubts were earlier expressed[26, 48, 108)] regarding the formation of such mixed adducts it is now, more or less, well-established that such mixed adducts do contribute to synergism. However, for thorium, for a given HA-B combination, three such mixed adducts, viz., $Th(NO_3)(TTA)_3 \cdot aTBP$, $Th(NO_3)_2(TTA)_2 \cdot bTBP$, and $Th(NO_3)_3(TTA) \cdot cTBP$ and one ternary adduct $Th(TTA)_4 \cdot B$ must be contributing to synergism and it is difficult to establish the composition of all these species, when they are present together, by extraction methods alone. Hence more work is necessary to establish the number and the nature of these mixed adducts in any given HA-B system. The reported values of some equilibrium constants involving such species[7, 94, 100)] need further confirmation.

The equilibrium constants for the equilibria involving the formation of $Th(TTA)_4 \cdot B$ are summarised in Table 11. The extraction of thorium by TBP in presence of different betadiketones was studied[109)] and while no synergism was observed for HAA, with others the extraction order HBA < HDFM < HTFA was observed, the species extracted being $ThA_4 \cdot TBP$ in all the systems.

Synergism was also observed[110)] in the extraction of thorium by HTTA in presence of bipyridyl, a bidentate neutral donor. However, the nature of the species extracted has not been investigated.

Synergism in the extraction of thorium with HIPT-B mixtures was studied[48, 111)]. The extraction of $Th(IPT)_4 \cdot B$ in these systems was well established. The formation of earlier

Table 11. Equilibrium constant data for the extraction of thorium by chelating acid – neutral donor combinations

Chelating acid	Neutral donor	Diluent	Aqueous medium	log K_A	log $K_{4,1}$	log $\beta^0_{4,1}$	Ref.
HTTA	TOPO	Cyclohexane	0.1 M HCl	–	10.45	–	9
HTTA	TBP	Cyclohexane	0.1 M HCl	–	7.85	–	9
HTTA	TPP	Cyclohexane	0.1 M HCl	–	5.54	–	9
HTTA	TBP	CCl_4	1 M $NaClO_4$	–	–	5.18	48
HTTA	TBP	$CHCl_3$	1 M $NaClO_4$	–	–	3.30	48
HTTA	TBP	$CHCl_3$	0.1 M $HClO_4$	1.00	4.24	3.24	103
HTTA	TBP	Cyclohexane	0.1 M $HClO_4$	2.73	9.80	6.07	103
HTTA	TBP	Benzene	0.1 M $HClO_4$	2.25	6.88	4.63	102
HTTA	TBP	Benzene	1 M $NaClO_4$	1.58	6.17	4.59	103
HTTA	TOPO	Benzene	0.1 M $HClO_4$	2.25	9.75	7.50	104
HTTA	TPPO	Benzene	0.1 M $HClO_4$	2.25	8.50	6.25	104
HTTA	DBBP	Benzene	0.1 M $HClO_4$	2.25	7.92	5.67	104
HTTA	DOSO	Benzene	0.1 M $HClO_4$	2.25	7.27	5.02	105
HTTA	TBTP	Benzene	0.1 M $HClO_4$	2.25	5.55	3.30	104
HIPT	TBP	CCl_4	1 M $NaClO_4$	–	–	1.7	48
HIPT	TBP	$CHCl_3$	1 M $NaClO_4$	–	–	1.2	48
HIPT	–	Benzene	0.1 M $HClO_4$	5.58	–	–	111
HIPT	TOPO	Benzene	0.1 M HCl	–	–	4.00	111
HIPT	TPPO	Benzene	0.1 M HCl	–	–	2.89	111
HIPT	DBBP	Benzene	0.1 M HCl	–	–	2.68	111
HIPT	DOSO	Benzene	0.1 M HCl	–	–	2.23	111
HIPT	TBP	Benzene	0.1 M HCl	–	–	1.78	111
HIPT	TBP	CCl_4	0.1 M HCl	–	–	2.60	111
HIPT	TBP	$CHCl_3$	0.1 M HCl	–	–	0	111

reported[48)] $Th(IPT)_4 \cdot 2\,B$ was shown[111)] to be incorrect and instead the presence of mixed adducts, probably $Th(IPT)_3(B)_j^+L^-$, were suggested[111)] to be present under certain experimental conditions. The values of the equilibrium constants for $Th(IPT)_4 \cdot B$ are included in Table 11.

Synergism was also reported[112)] in the extraction of thorium with mixtures of HDMPP and TBP.

Thorium is known[113)] to get extracted by oxine and several 1 : 1 neutral donor adducts of thorium oxinate were synthesised[114)]. Thus, mixtures of oxine and neutral extractants should provide a good combination for the synergic extraction of thorium.

It was reported[115, 116)] that by the use of MIBK or n-BA as diluents, the extraction of thorium by cupferron can be accomplished at a lower pH, which suggests that cupferron and neutral extractants can be good partners for the synergic extraction of thorium.

(b) Uranium(IV)

Earlier attempts[7)] on synergic extraction of U(IV) were unsuccessful as there was difficulty in preventing the oxidation of U(IV) to U(VI) during extraction. An attempt[117)] to prepare the neutral donor adducts of U(IV) betadiketonates did not succeed and hence it was concluded that such adducts are not possible with U(IV) due to its inability to reach nona coordination.

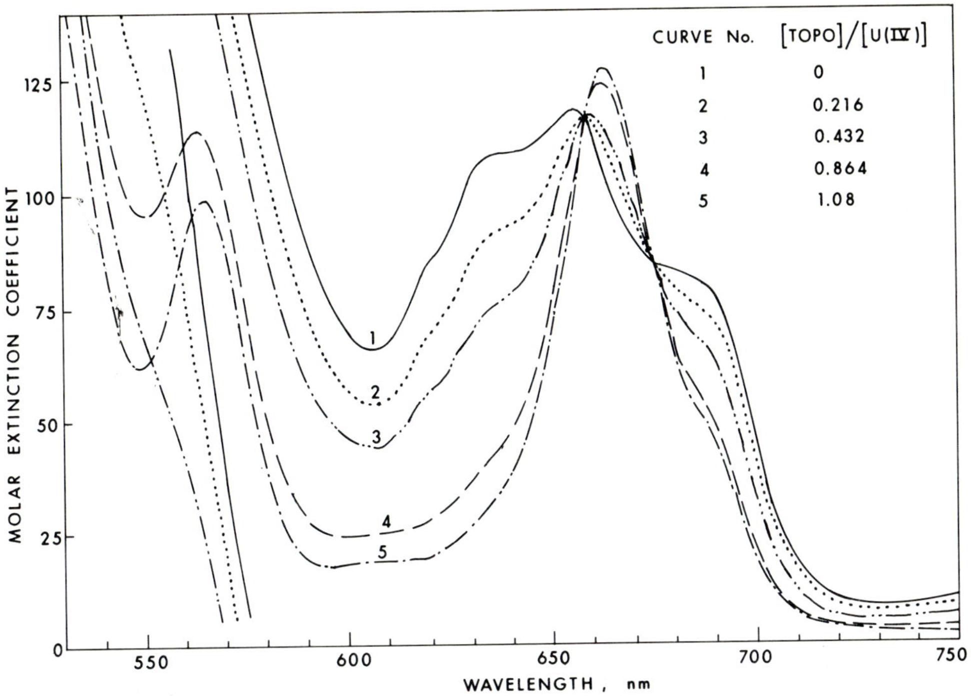

Fig. 3. Absorption spectra of U(IV) in excess HTTA as a function of different amounts of TOPO. (Reprinted from the report BARC-1007 (1979) by courtesy of the authors)

The extraction of HA-TBP combinations was studied[118)], taking special precautions to prevent the oxidation of U(IV), and it was suggested that $UA_4 \cdot 2$ TBP is responsible for the synergism. However, the subsequent spectrophotometric work[30, 119–122)] did not find any evidence for the formation of adducts other than $UA_4 \cdot B$, for these systems.

From spectrophotometric studies[30, 119–122)] $U(TTA)_4$ was shown to form 1:1 adducts with neutral donors of widely differing basicities. The absorption spectrum of $U(TTA)_4$ in benzene in presence of different amounts of TOPO is shown in Fig. 3, the isosbestic points of which suggest the formation of only one adduct in the system. The adduct formation constants are summarized in Table 12. The results of a recent nmr study[122)] are also included in this table. The difference between the nmr and the absorption spectral values remains to be explained. It may be connected with the diluent acetone which itself is a neutral donor[120)] of relatively poor basicity.

(c) Neptunium(IV)

From studies on the extraction of Np(IV) by HTTA-B mixtures from nitric acid medium, it was concluded[7, 123)] that mixed adducts involving nitrate ion participation get extracted. This conclusion was also substantiated with absorption spectral data[123)]. However, when the aqueous medium was changed to perchloric acid[105, 123–127)] mixed adduct extraction was very much suppressed and the synergism could be mainly attributed to the formation

Table 12. Equilibrium constant data for the adducts of $U(TTA)_4$; diluent – benzene; log K_A = 5.42 (for 1.0 M perchloric acid as aq. medium)

Neutral	log $K_{4,1}$[a]	log $\beta^0_{4,1}$	Ref.
TOPO	11.65	6.23	30
TBPO	11.55	6.13	120
TPPO	10.14	4.72	119
DBBP	9.46	4.04	30
DOSO	9.06	3.64	121
DHSO	9.03	3.61	121
DMSO	8.65	3.23	121
DBSO	8.52	3.10	121
TBP	8.46	3.04	119
TIOTP	6.70	1.28	30, 120
MIBK	5.32	−0.10	119, 120
TOPO[b]	–	3.98	120
TEP	–	3.02	122[c, d]
TEP	–	3.35	122[d]
DEA	–	2.55	122[c]
DEA	–	2.77	122[d]

[a] estimated values, [b] diluent – chloroform, [c] nmr data, [d] diluent – acetone

Table 13. Equilibrium constant data for the adducts of $Np(TTA)_4$; aqueous medium – 1.0 M perchloric acid; diluent – benzene; log K_A = 5.67

Neutral donor	log $K_{4,1}$	log $\beta^0_{4,1}$	Ref.
TOPO	11.33	5.66	126
TPPO	9.92	4.25	128
TPPO	–	4.27	128[a]
DBBP	9.33	3.66	127
DBBP	–	3.67	128[a]
DOSO	9.15	3.48	105
DOSO	–	3.33	128[a]
TBP	8.42	2.75	123
TBP	5.88	1.62	125[b]
TBP	10.45	4.31	124[c]

[a] Spectrophotometry, [b] diluent – chloroform, log K_A = 4.26, [c] diluent – cyclohexane, log K_A = 6.14

of $Np(TTA)_4 \cdot B$. When highly basic donors such as TOPO were used, however, additional species, which could be mixed adducts, were suggested[126)] to contribute to synergism. Mixed species, which may be ion-pairs, were also suggested[125)] in the extraction of Np(IV) by chloroform solutions of HTTA and TBP. The 1:1 adduct formation between $Np(TTA)_4$ and different neutral donors was independently substantiated by spectrophotometry[128)]. The absorption spectrum of $Np(TTA)_4$ and its changes in benzene as a function of a neutral donor concentration are shown in Fig. 4, which suggests a single adduct formation. The equilibrium constant data are given in Table 13. The good agreement in the extraction and spectrophotometric data further supports the conclusion that the said adduct is responsible for synergism.

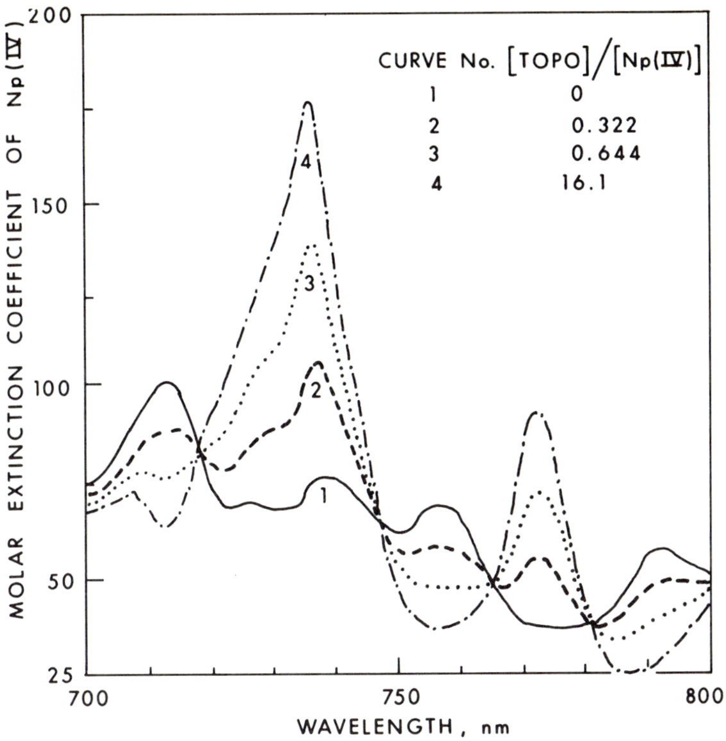

Fig. 4. Absorption spectra of Np(IV) in excess HTTA as a function of different amounts of TOPO. (Reprinted from Ref. 128 by courtesy of Indian Academy of Sciences, Bangalore)

(d) Plutonium(IV)

The extraction of Pu(IV) by HTTA-B mixtures was well studied[7, 24, 25, 52, 105, 127, 129, 130]. When nitric acid was used as the aqueous medium of extraction[7, 24, 25], mixed adducts involving nitrate ion participation were reported to get extracted causing synergism. In a recent investigation[24, 25] these adducts were identified to be $Pu(NO_3)_3(TTA) \cdot 2\,B$, $Pu(NO_3)_2(TTA)_2 \cdot B$ and $Pu(NO_3)(TTA)_3 \cdot B$ with the equilibrium constant data given in Table 14. When the aqueous medium employed was perchloric acid the 1:1 adduct formation, $Pu(TTA)_4 \cdot B$, was found to dominate[52, 105, 127, 129, 130]. However, mixed adduct formation involving perchlorate ion participation was also suggested[126]. The $Pu(TTA)_4 \cdot B$ formation was independently confirmed[52, 129, 130] using spectrophotometry. The isosbestic points in the spectra of $Pu(TTA)_4$ in benzene in presence of different amounts of a neutral donor, shown in Fig. 5, suggest the adduct formation. The equilibrium constant data for these systems are summarised in Table 15. The fact that the adduct formation constants obtained from extraction data are generally higher than the spectrophotometric data suggests that some more additional species are involved in the extraction, which was suspected[126].

Table 14. Equilibria and their stoichiometric constants (K) involved in the synergistic extraction of Pu(IV) from nitric acid medium by solutions of HTTA and some neutral organophosphorous donors (B)[24, 25]. Aqueous medium = 1.0 M nitric acid; diluent – benzene; Temp. – 25 °C

Equilibrium	Log K with the neutral donor (B)		
	TBP	DBBP	TOPO
$Pu^{4+} + 4NO_3^- + 2B_{(0)} \rightleftharpoons Pu(NO_3)_4 \cdot 2B_{(o)}$	1.95 ± 0.02	4.66 ± 0.03	10.08 ± 0.04
$Pu^{4+} + 3NO_3^- + HTTA_{(o)} + 2B_{(o)} \rightleftharpoons Pu(NO_3)_3(TTA) \cdot 2B_{(o)} + H^+$	5.21 ± 0.03	7.29 ± 0.05	12.25 ± 0.06
$Pu^{4+} + 2NO_3^- + 2HTTA_{(o)} + B_{(o)} \rightleftharpoons Pu(NO_3)_2(TTA)_2 \cdot B_{(o)} + 2H^+$	5.67 ± 0.12	7.13 ± 0.09	9.74 ± 0.10
$Pu^{4+} + NO_3^- + 3HTTA_{(o)} + B_{(o)} \rightleftharpoons Pu(NO_3)(TTA)_3 \cdot B_{(o)} + 3H^+$	8.33 ± 0.03	9.37 ± 0.04	11.78 ± 0.06
$Pu^{4+} + 4HTTA_{(o)} + B_{(o)} \rightleftharpoons Pu(TTA)_4 \cdot B_{(o)} + 4H^+$	9.08	10.18	12.27
$H^+ + NO_3^- + B_{(o)} \rightleftharpoons HNO_3 \cdot B_{(o)}$	−0.80	− 0.60	0.83
$Pu(NO_3)_4 \cdot 2B_{(o)} + HTTA_{(o)} + B_{(o)} \rightleftharpoons Pu(NO_3)_3(TTA) \cdot 2B_{(o)} + HNO_3 \cdot B_{(o)}$	2.46	2.03	3.00
$Pu(NO_3)_4 \cdot 2B_{(o)} + 2HTTA_{(o)} + B_{(o)} \rightleftharpoons Pu(NO_3)_2(TTA)_2 \cdot B_{(o)} + 2HNO_3 \cdot B_{(o)}$	2.12	1.27	1.32
$Pu(NO_3)_4 \cdot 2B_{(o)} + 3HTTA_{(o)} + 2B_{(o)} \rightleftharpoons Pu(NO_3)(TTA)_3 \cdot B_{(o)} + 3HNO_3 \cdot B_{(o)}$	3.98	1.91	4.19
$Pu(NO_3)_4 \cdot 2B_{(o)} + 4HTTA_{(o)} + 3B_{(o)} \rightleftharpoons Pu(TTA)_4 \cdot B_{(o)} + 4HNO_3 \cdot B_{(o)}$	3.93	3.12	5.51

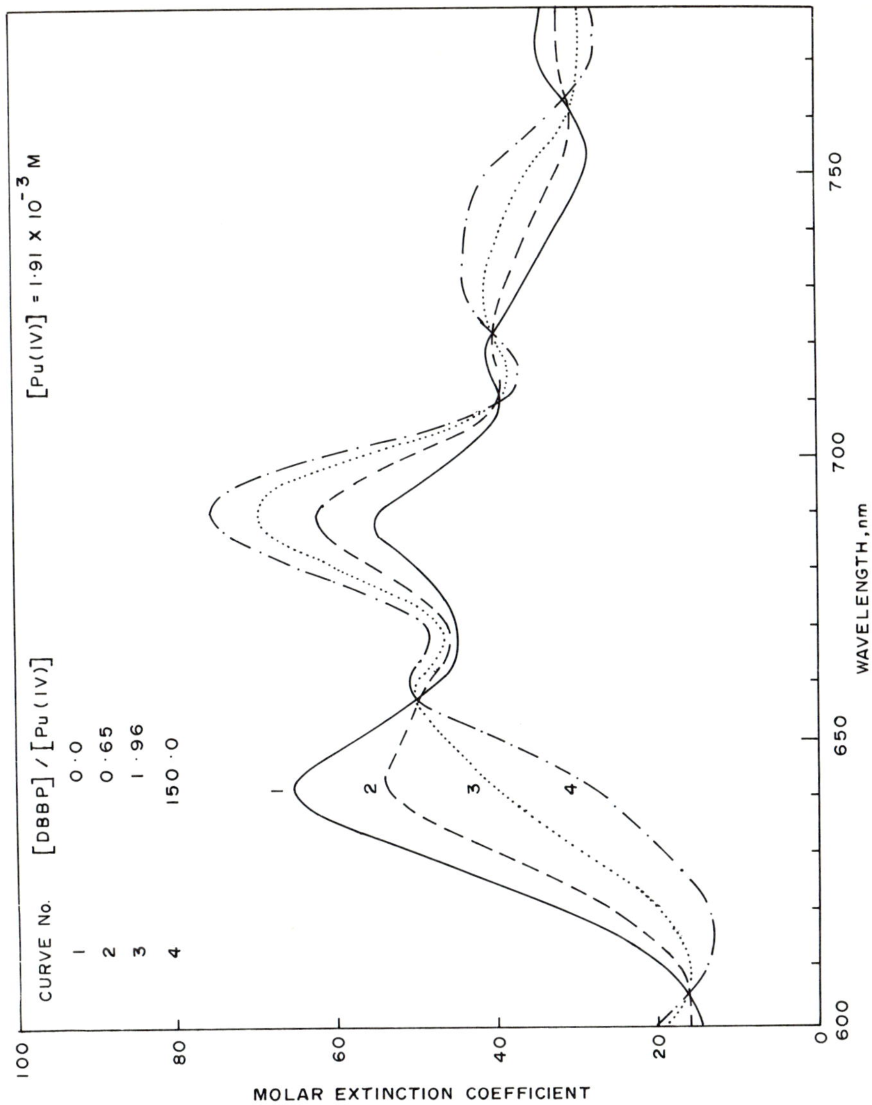

Fig. 5. Absorption spectra of Pu(IV) in excess HTTA as a function of different amounts of DBBP. (Reprinted from Ref. 129 by courtesy of Marcel Dekker Inc., New York)

Table 15. Equilibrium constant data for the adducts of $Pu(TTA)_4$ aqueous medium – 1.0 M perchloric acid; diluent – benzene; log K_A = 7.34

Neutral donor	log $K_{4,1}$	log $\beta^0_{4,1}$	Ref.
TOPO	12.29	4.95	126
TPPO	10.79	3.45	130
	–	3.04	130[a]
DBBP	10.17	2.83	129
	–	2.45	130[a]
	12.04	4.30	129[b]
DOSO	9.81	2.47	105
	–	2.31	130[a]
TBP	9.06	1.72	52
	–	1.83	52, 120[a]
	11.11	3.37	129[b]

[a] Spectrophotometry, [b] diluent – cyclohexane, log K_A = 7.74

III. Trivalent Ions

That mixtures of HTTA and TBP extract to a greater extent than either reagent used alone was first demonstrated for trivalent metal ions[3]. A large number of workers[8–11, 51, 131–141] have investigated the extraction of some trivalent actinide ions by HTTA in presence of different neutral extractants. A majority of them agree that the simple ternary adducts $M(TTA)_3 \cdot B$ and $M(TTA)_3 \cdot 2\,B$ are responsible for the observed synergism. The 1 : 1 adduct was, however, not noticed in the earlier work[8–11]. Those who used nitrate[8, 51] or thiocyanate[137] as the aqueous extraction media, however, reported that the extraction of different quaternary adducts, involving the aqueous anion participation, are also responsible for the synergism. The ion-pairs such as $M(TTA)_2(TOPO)_5^+$ ClO_4^-, reported for lanthanides[106, 107] were, however, considered by none in the corresponding actinide systems. In view of the possibility that different types of species, and that too a large number of them, are likely to be involved in synergism it is a difficult job to assign compositions to the extracted species and to derive the extraction constants. Unless special efforts are made to limit the number of the extracted species, by a proper choice of the experimental conditions, and different methods are used to confirm the nature of the extracted species, the results would be erroneous. This appears to be the case with a large data reported for trivalent actinide ions and the equilibrium constants are derived on the assumption that such and such species are involved in the extraction. More work is needed to establish these observations. However, the equilibrium constant data reported would be useful in assessing the extent of synergism. Such data reported for the simple ternary adduct extraction are summarized in Table 16. The previous summaries[12, 15,19, 142] where many more constants were derived from the earlier data[8–10] were excluded from this table since those involve the incorrect assumptions that the value of K_A is independent of the diluent used and the aqueous medium employed for extraction.

The data for Pu(III)[8] are approximate estimates since the authors faced the problem of maintaining the trivalent oxidation state of plutonium.

With some donors DBE, DIBE and DPS, the 1:4 metal chelate-neutral donor adducts were also reported[134] to be extracted. Such a conclusion could be incorrect as a result of neglecting the extraction of either ion-pairs or quaternary adducts.

It is seen from the data[134] that the ionic strength of the aqueous medium influences the adduct stabilities. As the adduct formation equilibrium is an exclusive organic phase equilibrium (see Eq. 17) it is not expected to be influenced by the aqueous phase conditions. Thus, these data clearly indicate that some species other than what were assumed seemed to be involved in the extraction. A similar study with the other systems would be highly useful in confirming the reported adduct compositions.

The observation that $Am(BTA)_3$ forms stronger adducts with MIBK than $Am(TTA)_3$[133] is also not expected[12].

The composition of the extracted adducts was reported[10, 138] to remain unaltered with change of diluents used, though the relative extractabilities differed very much. The equilibrium constant data are given in Table 17. Taking only one K_A value, for all the diluents[10], for the calculation of the other constants is not justified. Apart from this it needs to be carefully seen how far the adduct composition remains unaltered with diluent change because diluents were shown to change the adduct composition both with tetravalent actinide ions[111, 125, 130] as well as trivalent lanthanide ions[106, 107].

The extraction studies from nitrate media[8, 51, 141] suggest that different nitrate ion bearing species get extracted. While the earlier work[8, 51] suggested the extracted species as $M(NO_3)(TTA)_2 \cdot 2\,B$, a recent work[141] preferred the involvement of three species $M(NO_3)_2(TTA) \cdot 3\,B$, $M(NO_3)(TTA)_2 \cdot B$ and $M(TTA)_3 \cdot 2\,B$. The extraction from thiocyanate medium suggested[137] that both ternary as well as quaternary adducts do participate in synergism. More work is necessary to firmly establish the nature of these species.

The extraction of Am(III) by the chelating acids HCF or HO_x in presence of TPPO was studied[138] and the results are given in Table 18. While $Am(CF)_3$ was found to form two adducts $Am(Ox)_3$ was found to form only the first adduct. This may be due to the stronger complexing of Am(III) by oxine, which makes the residual coordination unsaturation in the metal chelate, much less. The steric hindrance could also play a role in preventing the second adduct formation.

The extraction of some trivalent actinide ions by HPMBP-B mixtures was reported[143–145]. The extraction constants obtained for different diluents[143] are given in Table 19. The constants were shown to decrease with increasing dielectric constant of the diluent. An additional mixed adduct, $M(NO_3)(PMBP)_2 \cdot 2\,TOPO$ participation was also considered when the neutral donor used was TOPO. The involvement of two ternary adducts in these systems was reported in the subsequent work[144, 145] and the equilibrium constant data are given in Table 20.

Table 16. Equilibrium constant data for the extraction of trivalent actinide ions by HTTA-B mixtures

M	B	Diluent	Aqueous	$-\log K_A$	$\log K_{3,1}$	$\log K_{3,2}$	$\log \beta^0_{3,1}$	$\log \beta^0_{3,2}$	Ref.
Pu(III)	TBP	Cyclohexane	dil. HNO_3	4.7	–	5.30	–	10.0	8
Am(III)	TBP	Cyclohexane	dil. HNO_3	6.62	–	3.41	–	10.03	8
	TOPO	Benzene	dil. HCl	–	–	2.51	–	–	9
	TPPO	Benzene	dil. HCl	–	–	0.85	–	–	9
	DBBP	Benzene	dil. HCl	–	–	0.51	–	–	9
	TBP	Benzene	dil. HCl	–	–	−0.96	–	–	9
	TPP	Benzene	dil. HCl	–	–	−4.63	–	–	9
	HTTA					–	0.75	–	131
	MIBK	CCl_4	1 M $NaClO_4$	8.88			1.80	2.40	132
	TBP	CCl_4	1 M $NaClO_4$	8.88			5.06	8.89	132
	DBSO	CCl_4	1 M $NaClO_4$	8.88			4.97	8.48	132
	MIBK	$CHCl_3$	1 M NH_4ClO_4				1.18	1.56	133
	MIBK	$CHCl_3$	1 M NH_4ClO_4				1.31	1.68	133[a]
	MIBK	$CHCl_3$	0.05 M NH_4ClO_4	9.13			0.8	1.1	134
	MIBK	$CHCl_3$	0.5 M NH_4ClO_4				1.05	1.4	134
	MIBK	$CHCl_3$	1.0 M NH_4ClO_4				1.15	1.5	134
	MIBK	$CHCl_3$	0.1 M NH_4ClO_4	9.13			0.9	1.2	134
	DBE	$CHCl_3$	0.1 M NH_4ClO_4	9.13			0.27	0.54	134
	DIBE	$CHCl_3$	0.1 M NH_4ClO_4	9.13			0.18	0.36	134
	DPE	$CHCl_3$	0.1 M NH_4ClO_4	9.13			0.28	–	134
	NBu	$CHCl_3$	0.1 M NH_4ClO_4	9.13			−0.5	− 0.26	134
	NBe	$CHCl_3$	0.1 M NH_4ClO_4	9.13			−0.4	− 0.22	134
	DPS	$CHCl_3$	0.1 M NH_4ClO_4	9.13			−0.22	− 0.12	134
	DPSe	$CHCl_3$	0.1 M NH_4ClO_4	9.13			0.13	–	134
	DBS	$CHCl_3$	0.1 M NH_4ClO_4	9.13			0.14	0.28	134
	DBSO	$CHCl_3$	0.1 M NH_4ClO_4	9.13			3.17	4.78	134
	$DBSO_2$	$CHCl_3$	0.1 M NH_4ClO_4	9.13			1.08	1.36	134
	D-n-PSO	Benzene	1.0 M $NaClO_4$	7.83	–	−0.14	–	7.69	136
	TBP	Xylene	0.005 M ClHAC	7.27	−3.08	0.18	4.19	7.45	139
	DOSO	Xylene	0.005 M ClHAC	7.27	−3.18	0.29	4.09	7.56	140
	TOPO	Xylene	0.005 M ClHAC	7.27	−0.87	3.85	6.40	11.12	140

Cm(III)	TOPO	Benzene	dil. HCl	–	–	2.83	–	–	9
	TBP	Benzene	dil. HCl	–	–	−0.70	–	–	9
	TBP	Xylene	0.005 M ClHAC	8.22	−3.76	−0.35	4.47	7.88	139
	DOSO	Xylene	0.005 M ClHAC	8.22	−3.78	0.07	4.46	8.16	140
	TOPO	Xylene	0.005 M ClHAC	8.22	−1.49	3.32	6.75	11.56	140
Bk(III)	TBP	Xylene	0.005 M ClHAC	6.12	−2.03	0.92	4.09	7.04	139
	DOSO	Xylene	0.005 M ClHAC	6.12	−2.51	1.14	3.61	7.26	140
	TOPO	Xylene	0.005 M ClHAC	6.12	0.09	4.31	6.20	10.43	140
Cf(III)	TBP	Xylene	0.005 M ClHAC	7.29	−2.55	0.31	4.73	7.60	139
	DOSO	Xylene	0.005 M ClHAC	7.29	−2.86	−0.01	4.43	7.27	140
	TOPO	Xylene	0.005 M ClHAC	7.29	−0.48	3.37	6.81	10.66	140

[a] The chelating used in this case is HBTA instead of HTTA

Table 17. Adduct formation constants of $Am(TTA)_3$ with neutral donors[10, 138] log K_A = − 8.89 was used for the data with TPPO and TOPO[138] and log K_A = − 6.8 was used for the data with TBP[10]

Diluent	TBP	TBPO		TOPO	
	log $\beta^0_{3,2}$	log $\beta^0_{3,1}$	log $\beta^0_{3,2}$	log $\beta^0_{3,1}$	log $\beta^0_{3,2}$
n-hexane	7.83	7.56	13.72	9.32	15.68
n-heptane	–	7.42	13.44	9.04	15.12
Cyclohexane	9.23	7.18	12.96	9.12	15.28
Isopropyl-benzene	–	6.52	10.88	7.96	13.12
CCl_4	7.23	6.67	11.34	7.98	13.09
Toluene	–	6.96	10.96	7.46	12.32
Benzene	5.8	6.74	10.88	7.57	12.18
$CHCl_3$	3.5	4.23	6.46	5.37	7.94
Chlorobenzene	–	6.34	10.32	7.28	12.20
Methylenechloride	–	5.09	8.48	6.15	9.83
O-dichloro-benzene	–	6.02	10.34	7.40	12.56
Bromoform	–	4.72	7.75	6.20	9.80

Table 18. Adduct formation constants of americium chelates with TPPO[138]

Diluent	HCF		HO_x
	log $\beta^0_{3,1}$	log $\beta^0_{3,2}$	log $\beta^0_{3,1}$
Isopropyl-benzene	4.74	8.08	3.62
CCl_4	4.34	7.90	3.76
Toluene	4.17	7.64	3.40
Benzene	4.22	7.56	3.15
Chloroform	3.94	6.32	2.34
Chlorobenzene	4.35	7.30	3.01
Methylene-chloride	4.21	6.74	2.79
O-dichloro-benzene	4.33	7.06	2.88

Table 19. Equilibrium constant data for the extraction of Am(III) by HPMBP-TBP mixtures[143]. Aqueous medium – 0.1 M HNO_3

Diluent	Dielectric constant	log $K_{3,2}$
n-hexane	1.890	5.18
n-heptane	1.924	4.99
n-octane	1.948	4.84
Cyclohexane	2.023	4.77
CCl_4	2.238	2.23
Toluene	2.379	1.63
Benzene	2.284	1.35

Table 20. Equilibrium constant data for the extraction of trivalent actinide ions by HPMBP-B mixtures

M	B	Diluent	Aqueous medium	$-\log K_A$	$\log K_{3,1}$	$\log K_{3,2}$	$\log \beta^0_{3,1}$	$\log \beta^0_{3,2}$	Ref.
Am(III)	MIBK	Chloroform	0.1 M NH_4ClO_4	4.43			0.59	0.96	144
	TBP	Chloroform	0.1 M NH_4ClO_4	4.43			0.77	2.00	144
	TBPO	Chloroform	0.1 M NH_4ClO_4	4.43			0.22	–	144
	TOPO	Chloroform	0.1 M NH_4ClO_4	4.43			0.9	7.6	144
	TPPO	Chloroform	0.1 M NH_4ClO_4	4.43			0.88	15	144
	TBP	Xylene	0.01 M HCl	2.73	−0.62	1.87	2.11	4.61	145
	TOPO	Xylene	0.01 M HCl	2.73	1.89	5.56	4.62	8.29	145
Cm(III)	TBP	Xylene	0.01 M HCl	2.64	−0.46	1.87	2.18	4.51	145
	TOPO	Xylene	0.01 M HCl	2.64	2.17	5.81	4.84	8.45	145
Bk(III)	TBP	Xylene	0.01 M HCl	1.23	4.43	–	3.26	–	145[a]
	TOPO	Xylene	0.01 M HCl	1.23	3.33	6.16	2.10	4.93	145
Cf(III)	TBP	Xylene	0.01 M HCl	1.49	0.52	2.32	2.02	3.81	145
	TOPO	Xylene	0.01 M HCl	1.49	3.02	6.23	4.51	7.72	145

[a] The species extracted was suggested to be $Bk(PMBP)_3 \cdot HPMBP \cdot TBP$

C. Miscellaneous Extractant Combinations

Apart from chelating acids several other organic acids such as organo phosphorus, carboxylic and sulphonic acids, are known to promote the extraction of metal ions, in combination with other extractants. Unlike chelating acids, these acids exist as polymers in non-polar diluents and extract metals by formation of solvated complexes, $MAn(HA)_p$. Another important feature of these polymeric acids is their strong tendency to interact with other extractants, such as neutral donors by hydrogen bonding, which is the main reason for the less marked synergism as well as the antagonism often observed with these combinations.

I. Combinations with Organophosphorus Acids

(a) Hexavalent Ions

That combinations of dialkyl phosphorus acids, 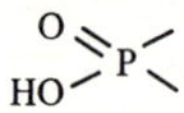, (HA) and neutral organophosphorus esters (B) give synergism in the extraction of U(VI) is the first well established study[2)] on synergic solvent extraction. This work clearly established that U(VI) extraction by a dialkyl phosphorus acid increases by the addition of a neutral ester in the order, phosphates < phosphonates < phosphinates < phosphine oxides, which is also the order of increasing basicity of the phosphoryl oxygen. Since then many more workers[146–158)] have studied these systems with a view to establish the nature of the species extracted. Different authors have expressed different opinions regarding the nature of the extracted species and a consensus seems to be rather difficult for the reasons described below.

The extraction of U(VI), at tracer concentrations, from dilute acid solutions, by dialkyl phosphorus acids, have been suggested to take place by four kinds of equilibria, by different authors. These are

$$UO_2^{2+} + 2\,HA_{(o)} \rightleftharpoons UO_2A_{2(o)} + 2\,H^+ \tag{28}$$

$$UO_2^{2+} + 2\,(HA)_{2(o)} \rightleftharpoons UO_2(HA_2)_{2(o)} + 2\,H^+ \tag{29}$$

$$UO_2^{2+} + 2\,(HA)_{2(o)} \rightleftharpoons UO_2A_2 \cdot 2\,HA_{(o)} + 2\,H^+ \tag{30}$$

and

$$UO_2^{2+} + 2\,(HA)_{2(o)} \rightleftharpoons UO_2A_3^-HA \cdot H^+_{(0)} + 2\,H^+ \tag{31}$$

In UO_2A_2 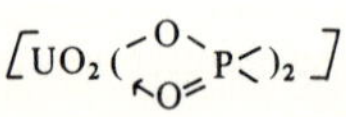 and $UO_2(HA_2)_2$ 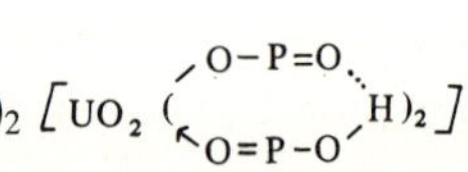 the coordination number of the atom is supposed to be six whereas in $UO_2A_2 \cdot 2\,HA$ and $UO_2A_3^-HA \cdot H^+$ it is eight. It is, however, not possible to distinguish between the above four equilibria using extraction methods alone since two molecules (either monomer or dimer) of the acid combine with the metal ion, in all the cases, liberating two hydrogen ions. Thus, all these differing opinions were carried forward in explaining the synergism in presence of the neutral additives.

Table 21. Equilibrium constants in the synergic extraction of U(VI) by mixtures of dialkylphosphoric acids and neutral donors

Acid	Neutral donor	Diluent	Aqueous medium	log K_{12}	log K_{121}	log β^0_{121}	Eqs.	Ref.
HDBP	TBP	Benzene	2 M HNO_3	4.69	9.64	4.95	(28, 32, 33)	149
HDAP	TBP	Benzene	2 M HNO_3	4.34	8.48	4.14	(28, 32, 33)	149
HDAP	DAMP	Benzene	2 M HNO_3	4.34	8.35	4.01	(28, 32, 33)	151
HDAP	TBPO	Benzene	2 M HNO_3	4.34	9.49	5.15	(28, 32, 33)	151
HAMP	TBP	Benzene	2 M HNO_3	3.15	6.35	3.20	(28, 32, 33)	151
HAMP	DAMP	Benzene	2 M HNO_3	3.15	6.87	3.72	(28, 32, 33)	151
HAMP	TBPO	Benzene	2 M HNO_3	3.15	8.23	5.08	(28, 32, 33)	151
HDEHP	TOPO	Kerosene	–	–	–	3.85	(35)	153
HDEHP	TBPO	Kerosene	0.1 M H_2SO_4	–	–	4.28	(35)	1
HDEHP	BDBP	Kerosene	0.1 M H_2SO_4	–	–	3.78	(35)	1
HDEHP	DBBP	Kerosene	0.1 M H_2SO_4	–	–	2.78	(35)	1
HDEHP	TBP	Kerosene	0.1 M H_2SO_4	–	–	2.18	(35)	1
HDEHP	TBP	Hexane	0.1 M H_2SO_4	–	–	1.85	(35)	1
HDEHP	TBP	CCl_4	0.1 M H_2SO_4	–	–	1.60	(35)	1
HDEHP	TBP	Benzene	0.1 M H_2SO_4	–	–	1.20	(35)	1
HDBP	TBP	CCl_4	0.1 M H_2SO_4	–	–	1.90	(35)	1
HDBP	TBP	CCl_4	0.1 M H_2SO_4	–	–	1.87	(38)	157
HDBP	TBP	Hexane	0.1 M H_2SO_4	–	–	2.82	(38)	157

Assuming Eq. (28) for U(VI) extraction by HA alone, the synergism was explained as due to the equilibrium,

$$UO_2^{2+} + 2\,HA_{(o)} + B_{(o)} \rightleftharpoons UO_2A_2 \cdot B_{(o)} + 2\,H^+ \qquad (32)$$

where the neutral donor was assumed to add to the metal chelate, increasing the coordination number of uranium from six to seven, as was observed earlier in the extraction by betadiketone – neutral donor mixtures, according to the equilibrium,

$$UO_2A_{2(o)} + B_{(o)} \rightleftharpoons UO_2A_2 \cdot B_{(o)} \qquad (33)$$

The equilibrium constants K_{12}, K_{121} and β^0_{121} of the equilibria represented by Eqs. (28), (32) and (33), respectively, reported[149, 151] are given in Table 21. In Eqs. (28) and (32) it was assumed that the acid HA to be present as a monomer in the diluents used. However, it is an established fact now that in most of the commonly used non-polar diluents these acids exist as dimers, $(HA)_2$ and hence the equilibria represented by Eqs. (28), (32) and (33) may not be considered seriously.

The addition mechanism was also considered by those[1] who assumed Eq. (29) to be responsible for U(VI) extraction by HA. Here the synergic extraction equilibrium becomes,

$$UO_2^{2+} + 2\,(HA)_{2(o)} + B_{(o)} \rightleftharpoons UO_2(HA_2)_2 \cdot B_{(o)} + 2\,H^+ \qquad (34)$$

involving the organic phase equilibrium,

$$UO_2(HA_2)_{2(o)} + B_{(o)} \rightleftharpoons UO_2(HA_2)_2 \cdot B_{(o)} \qquad (35)$$

In this adduct the B was thought to be adding to the metal chelate by hydrogen bonding[1, 2]. The equilibrium constants for the equilibrium represented by Eq. (35), reported[1] are included in Table 21.

Both the substitution equilibria represented by

$$UO_2A_2 \cdot 2\,HA_{(o)} + B_{(o)} \rightleftharpoons UO_2A_2 \cdot HA \cdot B_{(o)} + 1/2\,(HA)_{2(o)} \quad (36)$$

$$UO_2A_2 \cdot 2\,HA_{(o)} + 2\,B_{(o)} \rightleftharpoons UO_2A_2 \cdot 2\,B_{(o)} + (HA)_{2(o)} \quad (37)$$

and an addition equilibrium represented by

$$UO_2A_2 \cdot 2\,HA_{(o)} + B_{(o)} \rightleftharpoons UO_2A_2 \cdot 2\,HA \cdot B_{(o)} \quad (38)$$

were suggested from extraction studies involving HDBP and the neutral donors TBP and TOPO[147, 156, 157]. The substitution equilibria have found some support from ion exchange and infra red data[148, 150]. It was suggested that stronger neutral donors such as TOPO tend to form synergic species by substitution while the weaker ones like TBP form addition mechanism[147]. However, these different equilibria were arrived at based on computer calculations; slight changes in the mode of calculation changing not only the equilibrium constant numbers calculated but also the compositions of some species.

The equilibrium shown by Eq. (31) was suggested[150] based on infra red spectral data. In the presence of the neutral donors the substitution equilibrium,

$$UO_2A_3^-HAH^+_{(o)} + B_{(o)} \rightleftharpoons UO_2A_3^-BH^+_{(o)} + HA_{(o)} \quad (39)$$

and the addition equilibrium,

$$UO_2A_3^-HAH^+_{(o)} + B_{(o)} \rightleftharpoons UO_2A_3^-HA \cdot BH^+_{(o)} \quad (40)$$

were conceived[150, 152] and many extraction constants for the equilibrium,

$$UO_2^{2+} + 2\,(HA)_{2(o)} + B_{(o)} \rightleftharpoons UO_2A_3^-HA \cdot BH^+_{(o)} + 2\,H^+ \quad (41)$$

were reported[152].

The different conclusions regarding the nature of the species could not be confirmed from the data obtained at macroconcentrations of uranium since in the latter case, the metal chelate was found to be UO_2A_2[2] and no adduct, $UO_2A_2 \cdot B$ could ever be prepared.

Under certain experimental conditions, species such as $UO_2(HA)_2Cl \cdot 2\,B$[152] and $[UO_2A(NO_3)TBP]_2$[154] involving the participation of the aqueous phase bulk anion were also reported to be involved in the extraction.

When there exists a possibility of the participation of different types of equilibria, it is essential to consider all of them and no species should be excluded unless there is a strong evidence to do so. Thus it becomes a formidable task to find the type of species extracted in these systems. For these systems it may be of more practical utility to report the synergic coefficient values, under different experimental conditions, so that they may be useful in choosing the required extraction conditions.

More results on U(VI) extraction by different combinations of this type of systems have subsequently been published[158–167]. The synergic effect, in general, was found to increase with increasing acid strength of the organophosphorus acid[152] which is analogous to that observed with betadiketones. The synergic effects are, in general, found to be more in less polar diluents such as kerosene and decrease substantially even in chloroform[1,2,151].

Synergism was also observed in the extraction of Pu(VI)[146] and Am(VI)[168] by dialkylphosphorus acid – neutral donor combinations.

Addition of neutral donors to monoalkylphosphoric acids or phosphinic acids resulted only in antagonism of U(VI) extraction[2,169].

The extraction of U(VI) by both mono and dialkylphosphoric acids in presence of different secondary and tertiary amines was studied[146,170–172] using sulphuric acid as the aqueous medium. Regions of both synergism and antagonism were observed, for all these systems, depending on the aqueous phase sulphuric acid concentration and the molar ratio of amine/alkyl phosphoric acid in the organic phase. The equilibrium, governing the extraction with dialkylphosphoric acids in presence of TOA was suggested[170–172] to be

$$UO_2SO_4 + (HA)_{2(o)} + TOA \cdot H^+_{(o)} \rightleftharpoons UO_2SO_4(HA_2)^- TOA \cdot H^+_{(o)} + H^+ \tag{42}$$

and that with H_2MBP and TOA was suggested[172] to be

$$\begin{gathered} UO_2SO_4 + (H_2MBP)_{n(o)} + TOA \cdot H^+_{(o)} \\ \Updownarrow \\ UO_2SO_4(HMBP)(H_2MBP)^-_{(n-1)} TOA \cdot H^+_{(o)} + H^+ \end{gathered} \tag{43}$$

and both of these species in the organic phase were shown to be highly polymerized. The lowering of the aqueous sulphuric acid concentration resulted in a decrease of synergism, and finally giving rise to antagonism. This was attributed to the increasing interaction between the amine and the alkylphosphoric acid. These interesting systems are worth investigating by using other aqueous media with a view to ascertaining the formation of the ion-pairs suggested earlier (Eqs. 39, 41).

(b) Tetravalent Ions

Antagonism was noted in the extraction of thorium, by mixtures of dialkylphosphoric acids and neutral donors, from sulphate[2] and perchlorate[63] media, whereas a slight synergism at lower neutral donor concentration, followed by antagonism was noted[173] in the extraction from chloride medium. The extracted species responsible for synergism was supposed to be $Th(HA_2)_3Cl \cdot B$. On the other hand, mixtures of HDEHP and TBP were found[174] to greatly enhance the extraction of thorium from nitric acid medium. The species extracted was suggested to have the approximate composition, $Th(NO_3)(DEHP)_3 \cdot Th(NO_3)_4 \cdot 3\ TBP$.

Mixtures of H_2MEHP with neutral donors showed only antagonistic behaviour[169] in the extraction of thorium from dil. HCl and this was exploited in stripping thorium from the organic phase.

The absence of synergism in the extraction of U(IV) from aqueous phosphoric acid medium by HDEHP-TOPO was assumed[165] as due to the coordination saturation of U(IV) in its complex with HDEHP. However, H_2MEHP-B combinations have been used[175] for the extraction of U(IV) from dilute phosphoric acid where the synergism observed was inferred to be due to the extraction of $U(H_{8n-4})MEHP_{4n}) \cdot B$ where B = TOPO or DBBP.

By addition of TBP to H_2MEHP the extraction of Np(IV) from sulphuric acid by a toluene solution of H_2MEHP, could be lowered by a factor of $\sim 10^5$ [176]. This was attributed to the strong interaction between the reagents forming $(TBP)_2(H_2MEHP)_6$.

Addition of a small amount (5×10^{-3} M) of HDIDP was reported[177] to enhance the extraction of Pu(IV), from nitric acid, by 30% TBP. No species, responsible for this, has however been suggested.

(c) Trivalent Ions

In the extraction of Am(III) from 0.1 M nitric acid by HDBP, it was seen[178] that the use of diluents such as TBP or hexol gave poor extraction as compared with hexane, CCl_4 or $CHCl_3$. But when low concentrations of TBP were used synergism was found[168] in the extraction with HDEHP or HOPP. The synergism was explained as due to the involvement of the equilibrium,

$$Am^{3+} + 3\,(HA)_{2(o)} + B_{(o)} \rightleftharpoons Am(HA_2)_3 \cdot B_{(o)} + 3\,H^+ \tag{44}$$

as was explained earlier[173] in the extraction of Am(III) or Cm(III) from dil. HCl, by HDOPP-B mixtures. TBP was reported[178] to cause only antagonistic effect in the extraction of Am(III) by HDEHP from perchlorate, citrate or trinitriloacetate media at pH = 2.87. At low TBP concentrations, the species $AmA_3 \cdot$ HAyTBP was suggested to be extracted.

The extraction of Am(III) or Cm(III) by H_2MEHP from 0.25 M HCl with increasing concentrations of n-decanol or TBP showed[169] slight synergism followed by antagonism. The synergism was attributed to a partial adduct formation between the neutral donor and H_2MEHP which is having a replaceable hydrogen ion where as antagonism, due to complete adduct formation. The antagonism as measured by the depressent factor, for a fixed $[H_2A]$ and 0.1 M B was found to be 1.7 (n-decanol), 22 (TBP) and 1000 (TOPO).

Mixtures of HDEHP and H_2MEHP at a mole ratio of 2–3 enhanced[179] the extraction of Am(III) from nitric acid by a factor of more than 10^3, thus enabling its extraction at aqueous acid concentrations exceeding 1 M.

Mixtures of HDEHP (HA) and HDNNS (HS) were observed[180, 181] to give synergism in the extraction of Am(III) from 1.0 M perchloric acid, and the equilibrium involved was suggested to be

$$Am^{3+} + (HA)_{2(o)} + (HS)_{7(o)} \rightleftharpoons AmA_2H_6S_{7(o)} + 3\,H^+ \tag{45}$$

II. Combinations with Carboxylic Acids

(a) Uranium(VI)

In the extraction of U(VI) by carboxylic acid-neutral donor mixtures[13, 182–185], both synergism and antagonism were observed depending on the aqueous pH. High pH promotes dissociation of the acid and complex formation of the metal ion with the acid anion with subsequent extraction of the metal complex by solvation with the neutral donor. On the other hand, low pH promotes the extraction of the undissociated acid into the organic phase which then interacts with the neutral donor, thus virtually removing both the extractants from the system. The pH effect, thus, can be taken advantage in promoting or suppressing extraction, as desired using these mixtures. Some representative data[13] are given in Table 22. At a given pH, stronger acid (More dissociating) is seen to give higher synergism, as in the case with betadiketones. The unusually high extraction with salicylic acid was attributed to a different type of bonding with this acid. Such differences with salicylic acid were almost invariably observed in many known metal-ligand stability constant values[186]. The composition of the species extracted appears to be complicated due to polymerisation in the organic phase, though the species $UO_2A_2 \cdot 2B$[185] and $UO_2(ClO_4)A \cdot 2B$[184] were suggested to be extracted in some cases. The latter species was suggested in the extraction of U(VI) from 0.1 M $NaClO_4$, by mixtures of TOPO and benzoic acid, with an extraction constant value of $\log K = 7.00$ for the equilibrium,

$$UO_2^{2+} + ClO_4^- + HA_{(o)} + 2\,TOPO_{(o)} \rightleftharpoons UO_2(ClO_4)\,A \cdot 2\,TOPO_{(o)} + H^+ \quad . \tag{46}$$

Long chain carboxylic acid tertiary amine combinations were also found to give synergism for U(VI) extraction[60, 182, 187–189]. As expected both synergism and antagonism have been observed with these systems depending on the aqueous pH. The species extracted from these systems were reported to be either $UO_2A_2 \cdot TLA$, $UO_2A_2 \cdot HA \cdot TLA$ or $UO_2A_2 \cdot 2(HA \cdot TLA)$ depending on the acid or the diluent used and these compositions are similar to those observed in the extraction of U(VI) by tertiary amines from aqueous, carboxylic acids[190–192].

Table 22. Extraction data for U(VI) by 0.02 M TOPO in presence of 0.01 M of different carboxylic acids. Aqueous phase – 0.01 M HNO_3; diluent – CCl_4

Carboxylic acid	pK	Distribution ratio at pH =	
		3.85	4.90
phenylpropiolic	2.23	214	363
O-chlorobenzoic	2.92	105	223
Salicylic	2.97	365	693
Benzoic	4.17	21.8	93.9
trans-cinnamic	4.44	14.9	32.9
hydroxy-cinnamic	4.74	6.9	37.4
O-nitrobenzoic[a]	2.21	4.0	8.4

[a] 0.0015 M. Note: The extraction by TOPO alone at this pH values is negligible

(b) Tetravalent Ions

As an addition compound was reported to be formed[193] between thorium salicylate and an undissociated salicylic acid molecule, addition of neutral donors to salicylic acid or other carboxylic acids may give regions of synergism and antagonism with tetravalent ions also, as with U(VI). No study however has been reported on this.

In the extraction of Np(IV), from nitiric acid, by carboxylic acid-TLA mixtures, both synergism and antagonism were observed[60], as in the case of U(VI).

III. Combinations with Chelating Acids

(a) Oxycations

It has been known[194] that the oxine complex of U(VI) could be readily extracted from aqueous alkaline solutions into chloroform or MIBK in the presence of quaternary ammonium or arsonium cation from which it is clear that chelating agents which form anionic complexes with U(VI), can give synergism in combination with organic cation contributing extractants such as amines. However, feeble synergism was noted in the extraction of U(VI) from nitric acid[195] or perchloric acid[196] by HTTA and TOA. With these mixtures, however, a close pH control is essential since high aqueous acidity inhibits chelation and high pH makes the tertiary amine behave like a weak neutral donor.

Np(V) has shown much enhanced extraction by HTTA-TOMA mixtures[57]. The synergism was reported to be due to the equilibrium,

$$NpO_2^+ + TTA^- + 2(TOMAHTTA)_{(o)} \overset{K}{\rightleftharpoons} NpO_2(TTA)_3((TOMAH)_{2(o)} \qquad (47)$$

with log K = 2.3. By using a proper combination of the extractants log D > 2 could be realized, whereas with the individual extractants log D > −2 could not be achieved.

The effect of mixing two chelating acids, HTTA and HO_x, on the extraction of U(VI) was studied where synergism was observed[197]. Though the enhanced extraction was attributed to the mixed chelate formation, many species are likely to get extracted in these systems in view of the ability of the undissociated chelating acids to behave as neutral donors.

(b) Tetravalent Ions

The extraction of thorium by mixture of HTTA and TOA was studied[198] and the observed synergism was attributed to the equilibrium,

$$Th(TTA)_{4(o)} + TOAHCl_{(o)} \rightleftharpoons Th(TTA)_4 \cdot TOAHCl_{(o)} \qquad (48)$$

with log K = 2.42.

As thorium is known to form pentakis oxinates[193] or pentakis tropolonates[114], extraction of thorium by oxine or tropolones in combination with amines or other cation contributing extractants is expected. Such a combination was successfully employed in a method[199)] for the extractive photometric determination of thorium.

As pentakis tropolnates of U(IV) are also known[200)] the extraction chemistry of U(IV) is expected to be closely similar to that of thorium.

Mixtures of HIPT and HTTA were shown[201)] to give synergism in the extraction of thorium. In addition to the mixed chelates, several HIPT adducts of the mixed chelates were suggested to be responsible for the synergism.

(c) Trivalent Ions

Extraction of some trivalent actinide ions into mixtures of HTTA and amines was studied[131, 202)]. The order of the metal ion extraction with respect to the aqueous medium was reported[202)] to be chloride > bromide > thiocyanate > iodide. As the anions of the bulk aqueous medium compete with anionic metal complexes, in getting associated with the amine cation, the metal ion extraction is supposed to be least when the bulk anion has a preferential association with the amine cation. As the least hydrated iodide ion, being large, is preferred by the amine cation, it is logical that the metal complex is least extracted in that medium. The extracted species from 0.1 M KCl medium, were suggested[202)] to be the ion pairs $M(TTA_3)Cl^-ALIQ^+$ according to the equilibrium,

$$M^{3+} + 3\,HTTA_{(o)} + ALIQCl_{(o)} \rightleftharpoons M(TTA_3)Cl^-ALIQ^+_{(o)} + 3\,H^+ \qquad (49)$$

with $\log K = -0.66$ (Am), -0.78 (Cm) and -0.51 (Cf), for xylene as diluent. The adduct formation equilibrium,

$$M(TTA)_{3_{(o)}} \quad + \quad ALIQCl_{(o)} \overset{\beta^0}{\rightleftharpoons} M(TTA)_3Cl^-ALIQ^+_{(o)} \qquad (50)$$

with $\log \beta^0 = 7.27$ (Am), 7.13 (Cm) and 6.90 (Cf), was also reported. The decrease in the adduct stabilities with increasing atomic number of the metal ion followed an analogous behaviour with that of lanthanides.

IV. Combinations with Neutral Extractants

(a) Hexavalent Ions

Amines are known to extract U(VI), from nitrate media, as $UO_2(NO_3)_3^-$. If any neutral donor replaces one of the nitrate groups, the anionic nature of the species is destroyed and hence no synergism in combination with amines would be expected[203)]. Synergism was, in fact, not observed in the extraction of U(VI), from nitric acid, by amine-neutral donor combinations[204)]. However, if acids such as hydrochloric or sulphuric are employed as aqueous media, the anionic nature of the complexes $UO_2Cl_4^{2-}$ or $UO_2(SO_4)_3^{4-}$, even after replacing one Cl^- or SO_4^{2-} with neutral donors can be retained and hence synergism may be likely in combination with amines.

Synergism by mixing different neutral extractants was observed[205)] for U(VI) and this was attributed to the solvation by both the donors.

Neutral extractants in combination with HDNNS were observed[206] to give synergism in the extraction of U(VI), Np(VI) and Pu(VI).

(b) Tetravalent Ions

As tetravalent ions are extracted by amines as $M(NO_3)_6^{2-}$ synergism in combination with neutral extractants was expected[203] and it was in fact observed[207] in the extraction of thorium by mixtures of TDA with TBP or TOPO.

Synergism[208], antagonism[(209)] and neither of the two[210] was reported in the extraction of Pu(IV) from nitric acid by TBP-TLA mixtures. A systematic study is required to elaborate on these qualitative statements.

Synergism by mixtures of different neutral donors was observed in the extraction of thorium from nitrate media[101, 207] and chloride media[211] and that of Pu(IV) from nitric acid medium[212], and the mixed solvates were assumed to be responsible for it.

(c) Trivalent Ions

TOPO-TLA mixtures were found[208] to give synergism with Am(III) or Cm(III) when the aqueous medium contained lithium nitrate and antagonism with nitric acid alone. This could be due to the extraction of M(III) as anionic nitrate complex which is promoted by the presence of lithium nitrate. Others[61] also observed synergism with TOA-B, from nitrate solutions, and attributed it to the extraction of species such as $[M(NO_3)_5(B)_n^{2-}(TOA \cdot H^+)_2]$ or $[M(NO_3)_4(B)_n^{-}(TOA \cdot H^+)]$ with $n = 1$ or 2.

Mixtures of TBP and DBDECMP were found[212] to give synergism in the extraction of Am(III) from nitric acid.

D. Physico-Chemical Studies

The adduct formation between metal chelates and neutral donors is seen to be mainly responsible for the synergic extraction of metal ions by mixtures of chelating acids and neutral extractants. With a view to establish the composition and the nature of these adducts different physico chemical studies were undertaken for these systems. Regarding the site of bonding for the neutral donor, two possibilities were suggested; (1) the neutral donor is bound with the chelating ligands which were coordinated to the metal ion, and (2) the neutral donor is directly coordinated to the metal ion. The latter view was considered[15] along with three possibilities. They are that (1) the neutral donor is coordinated to the metal ion, without changing its coordination number, by making a bidentate chelating ligand, monodentate, (2) the neutral donor is coordinated to the metal ion, without changing its coordination number, by replacing a coordinated water molecule in the hydrated metal chelate, and (3) the neutral donor is coordinated to the metal ion by increasing its coordination number. The different studies undertaken to elucidate these aspects are given here.

I. Uranium(VI)

(a) Absorption Spectra

The adduct formation between $UO_2(TTA)_2$ with an undissociated molecule of HTTA[70] and a molecule each of different neutral donors[96] such as TBP, DBBP or TOPO were confirmed from visible absorption spectral work. The formation of adducts, other than of 1:1 composition in these systems, such as $UO_2(TTA)_2 \cdot 2$TOPO or $UO_2(TTA)_2 \cdot 3$TOPO and that of $UO_2(TTA)(NO_3) \cdot$ TBP which were reported earlier[5, 9, 10, 213, 214] were shown to be absent from the absorption spectral study[96].

(b) Thermodynamic Studies

The thermodynamic data for the adduct formation equilibrium between uranyl chelates and neutral donors (Eq. 17) were evaluated from the solvent extraction data[90, 91] obtained at different temperatures and these are given in Table 23. It is seen that these adducts are mainly enthalpy stabilized and this is expected when the neutral donor is directly coordinated to the metal ion. However, caution should be exercised in interpreting such data since the diluent used is known to influence thermodynamic parameters appreciably[215].

(c) Isolation of Adduct Compounds

Several neutral donor adducts of uranylbetadiketonates were isolated[89, 213, 214, 216–220]. That these adducts are monomers in benzene was verified for some of them from molecular weight determinations[89, 213, 214, 217]. Though it was suggested that water could accom-

Table 23. Thermodynamic data for the adduct formation of uranyl betadiketonates with some neutral donors[90, 91]. Equilibrium, $UO_2A_2 + B \rightleftharpoons UO_2A_2 \cdot B$; diluent – benzene

Neutral donor (B)	Betadiketone (HA)	$-\Delta G$ (kcal/mol)	$-\Delta H$ (kcal/mol)	ΔS (e.u.)
TOPO	HTTA	6.59	3.77	9.46
TBP	HTTA	5.28	5.21	0.24
DBSO	HTTA	4.76	4.90	-0.47
DPSO	HTTA	4.13	4.60	-1.58
DHSO	HTTA	5.48	3.52	6.60
DISO	HTTA	5.42	3.81	5.40
DOSO	HTTA	5.31	3.05	7.60
DPSO	HBTA	3.64	4.29	-2.20
DBSO	HBTA	4.27	3.84	1.44
DISO	HBTA	5.14	4.46	2.28
DHSO	HBTA	5.26	3.81	4.83
DOSO	HBTA	5.17	4.58	1.98
DHSO	HBA	4.16	5.03	-2.92
DISO	HBA	3.70	2.60	3.69
DOSO	HBA	3.90	3.07	2.78

pany some of these adducts in the extraction process[221], all the adduct compounds isolated were shown to be anhydrous. The different investigations carried out with these adducts are described below.

i) Infrared Spectra

From the observed absence of absorption in 3500 cm^{-1} region, it was suggested[89, 213, 214, 217–220] that all these adducts are anhydrous. From the changes in ir absorption spectra of neutral donor alone and that of its adduct with uranyl chelates, it was established that the neutral donor is directly coordinated to uranyl ion.

The splitting of the maxima, present around the 1600 cm^{-1} region, in the adducts of $UO_2(TTA)_2$ was interpreted[213] as due to the presence of some free (uncoordinated) chelating acid (TTA) in the adducts. This led to the conclusion[213] that one TTA group is bound to the metal as a monodentate ligand in the 1:1 adducts, making the uranium atom six-coordinated in them. By a similar interpretation, it was suggested[213] that both the TTA groups are bound to the metal as monodentate ligands in the adduct of the supposed composition, $UO_2(TTA)_2 \cdot 3\,TOPO$, making the uranium atom seven-coordinated in it. Since anhydrous $Cu(TTA)_2$ and $Zn(TTA)_2$, in which both TTA moities are bidentate, have shown two bands around 1600 cm^{-1}, like the above adducts, it was suggested[214] that the above interpretation was not correct and that there was no monodentate TTA in the above adducts. The splitting of the peak was observed only when the betadiketone involved was fluorinated and not when it was HAA or HDBM[89, 217–219] from which it was concluded that the split is due to the electronegative trifluoromethyl groups and not due to the monodentate TTA.

The infra red spectra of several Pu(VI) betadiketonate – neutral donor adducts were recorded[99] and it was concluded that they are identical with the anlogous uranyl complexes.

ii) Nmr Spectra

The nmr spectra of the $UO_2(TTA)_2$ adducts were studied[222] and only a single sharp line was observed, for the metal ring hydrogen, rather than the expected double line if one of the TTA were monodentate. The suggestion[222] that a single line might have been observed due to a rapid exchange of the two non-equivalent hydrogens, did not find any evidence in the subsequent work[89] which showed that the resonance was not altered by changing the temperature.

The fluorine resonance of the complexes $UO_2(TTA)_2 \cdot 2\,H_2O$, $UO_2(TTA)_2 \cdot TBP$ and $UO_2(TTA)_2 \cdot DPSO$ gave one line from which it was concluded[89] that these molecules are symmetric with equivalent trifluoromethyl groups.

The $UO_2(TTA)_2$ adducts with sulphoxides and TBPO have shown the presence of a doublet for the γ-CH proton and additional shoulder peaks for TTA protons from which it was concluded[220] that these complexes exist as a mixture of cis-trans isomers.

iii) X-ray Crystal Structures

During the course of thermal studies[223)] with the compound, $UO_2(HFA)_2 \cdot TMP$, three polymorphs of this complex were detected and the x-ray molecular structures of two of them were determined[224, 225)]. The x-ray molecular structures of $UO_2(TTA)_2 \cdot TOPO$[226)], $(UO_2(AA)_2 \cdot TPPO) \cdot C_6H_6$ and $UO_2(ClAA)_2 \cdot TPPO$[227)] were also studied. It was concluded from all these studies that the uranium atom adopts seven-coordinate pentagonal bi-pyramidal geometry, four oxygen atoms from two bidentate betadiketone groups and one oxygen from the neutral door, forming the pentagon. Similar structures were suggested for the adducts $UO_2(O_x)_2 \cdot HO_x$[228, 229)], $UO_2(AA)_2 \cdot B$[230, 231)] and $UO_2(AA)_2 \cdot HAA$[232)] or $UO_2(DBM)_2 \cdot H_2O$[233)] as also for $UO_2(HCOO)_2 \cdot H_2O$[234)].

iv) Melting Points

The various neutral donor adduct compounds of the uranyl chelates prepared[89, 213, 216–218)] are listed in Table 24, along with their melting points. One striking feature of these data is the adducts appear to have much lower melting points compared to the parent chelates and the higher the basicity of the neutral donor, the lesser is the melting point. This decrease in melting point is a reflection of the coordination saturation of the uranium atom. The exception to this rule is observed with some donors such as those containing phenyl groups which may be promoting association of the adducts probably involving pi-electrons.

Table 24. Melting points of the 1:1 adducts between metal chelates and neutral donors, °C.

Metal ion	Neutral donor	Chelating Acid						
		HTTA	HAA	HDBM	HBTA	HTFA	HBA	HHFA
UO_2^{2+}	–	212			156			
	TOPO	57	–	–	59	–	–	–
	TEHPO	60	–	–	–	–	–	–
	TPPO	238	–	–	–	–	–	–
	TBP	112	–	–	80	–	–	–
	DHSO	105	–	145	114	–	129	–
	DISO	147	–	174	140	98	150	–
	DBSO	128	220	300	144	–	185	–
	DPSO	166	170	228	152	–	190	–
	PO	181	–	238	183	144	212	160
	4MOPO	179	–	206	174	120	110	95
	4MPO	190	–	258	175	148	191	147
	4CPO	188	–	256	193	125	130	145
	4NPO	240	–	250	182	134	147	200
Th^{4+}	–	224	–	–	–	140	–	103
	TOPO	110	–	–	–	oil	–	oil
	TPPO	305	–	–	–	87	–	269
	TBP	178	–	–	–	oil	–	oil
	TPP	–	–	–	–	–	–	169
U^{4+}	–	250						
	TOPO	92						

The solubilities of these adducts, in non-polar solvents, are also expected to follow a similar behaviour, the stronger adducts dissolving more easily and giving higher concentrations in solution.

(d) Miscellaneous

Structural aspects of the actinide carboxylate complexes have been reviewed[235] from which the structures of many adducts that are likely to be involved in the extraction with carboxylic acid – neutral donor mixtures may be predicted.

There are practically no studies on the isolation and evaluation of the adducts extracted into organophosphorus acid combinations. However, ion exchange[148] and infrared measurements[150] were attempted to show that the synergic species is formed by the substitution of the neutral donor for an acid molecule rather than addition to the metal-acid complex. Infrared data have also suggested[150] that an ion-pair of the type $[UO_2A_3^-BH^+]$ could also be involved in synergism with the neutral donor – organophosphorus acid mixtures. Similar studies in these systems suggested[155] that the uranium atom is bonded to the $P \rightarrow O$ of HDEHP and not to TBP. However, the thermodynamic data obtained by the temperature variation method[155] as well as by calorimetry[162] appear to favour a direct bonding of TBP with the uranium ion.

(e) Conclusions

The suggestion that, in the neutral donor adducts of uranyl chelates, the neutral donor is bound with the chelating groups was made[19] based on the available extraction equilibrium constants. However, all the physico-chemical studies discussed above favour a direct bonding of the neutral donor to the uranium atom. The studies also confirm that the bidentate nature of the chelating ligands continues even after the adduct formation. Now, whether the neutral donor is replacing a water molecule that was coordinated to the metal chelate or is adding to an anhydrous metal chelate depends on whether the metal chelate is hydrated or anhydrous. While some of the studies tend to show that the bare chelate is hydrated, others, including the spectrophotometric work[96], suggest that at least $UO_2(TTA)_2$ is anhydrous in benzene and the neutral donor is coordinated to it by increasing the coordination number of uranium from six to seven. More work seem to be necessary to establish this aspect.

II. Tetravalent Ions

The high solubilities of some metal betadiketonates in polar solvents, as compared to non-polar solvents, was attributed[236] to the direct coordination betweeen the polar solvent and the central metal atom. On this basis, it was concluded[5] that tetravalent metal ion chelates are coordinatively saturated as they were reported to be readily soluble in non-polar solvents and only sparingly soluble in polar solvents. This conclusion would have been correct, but for the incorrect solubility data available to them.

Many adduct compounds of thorium chelates were prepared and the composition and melting points of some of them[117, 213] are included in Table 24. Attempts were made[213], using infrared work to show that one of the chelating ligands, in the adducts, is monodentate, thereby retaining the assumed maximum coordination number of eight for thorium. In the subsequent work[117] attempts to prepare such adducts for Th(IV), U(IV), Hf(IV) and Zr(IV) met with success only with Th(IV). This led to infer that the central metal atom takes part in the adduct formation by increasing the coordination number from eight in the chelates to nine in these adducts. Subsequent x-ray crystal structure studies with $Th(TTA)_4 \cdot TOPO$ confirmed[237] that Th(IV) has a coordination number of nine in it, the structure being a tricapped trigonal prism. Nine coordination for thorium was reported prior to this[238] in a similar adduct, $Th(TR)_4 \cdot DMF$, the structure of this compound being a monocapped square antiprism. Recent themodynamic studies[215] on the adduct formation between $Th(TTA)_4$ and TBP are also in agreement with the direct addition of TBP to $Th(TTA)_4$, without replacing any coordination position occupied by TTA groups. The water content measurements with the extraction of thorium by mixtures of HTTA and DIPY or PHEN[110] did not reveal the displacement of any water molecules, upon the addition of the neutral donor to the metal chelate, thus conclusively showing that adduct formation in this case occurs by addition of the neutral donor to the metal ion by increasing the coordination of the metal ion from eight in the chelate to nine in the adducts.

The kinetics of ligand exchange between a U(IV) betadiketonate and the free betadiketone was studied[239] using nmr and it was found that the exchange involves a ninth coordination site in the U(IV) chelate. Subsequently nmr was employed[122] to study the adduct formation between $U(TTA)_4$ and some neutral donors. The absorption spectral studies with the chelates of U(IV) and a number of neutral donors[30, 119–122] clearly sugested that the neutral donor addition involves an expansion in the coordination number of U(IV) from eight in the chelates to nine in the adducts. Though the earlier attempt[117] to prepare a U(IV) adduct did not succeed, the adduct $U(TTA)_4 \cdot TOPO$ was subsequently prepared[240] and was found to be isostructural with its thorium analogue, thus establishing its nine-coordination in these adducts.

The absorption spectral studies with $Np(TTA)_4$[128] and $Pu(TTA)_4$[52, 129, 130], in the presence of different neutral donors also suggest that adduct formation with these metal ions also is identical to that observed with Th(IV) and U(IV).

Some of the available data on the neutral donor adduct formation of $M(TTA)_4$ chelates are compared against the neutral donor basicities in Fig. 6. The parallel relationships observed between different metal ions suggest that the adducts involved are, most probably, isostructural. A similar relationship showing the diluent effect is given in Fig. 7 and a comparison between the stabilities of the $Th(TTA)_4$ adducts and $Th(IPT)_4$ adducts is shown in Fig. 8. Such correlations are of much practical value in estimating the stabilities of the neutral donor-metal chelate adducts from the data available for similar systems.

III. The Diluent Effect and Destruction of Synergism

The diluents employed influence significantly the extent of synergic extraction of metal ions by different combinations of extractants. Generally with relatively polar diluents such as chloroform synergic enhancement is much less than that observed with non-polar solvents like hexane or cyclohexane. With the former diluents the metal species extracted

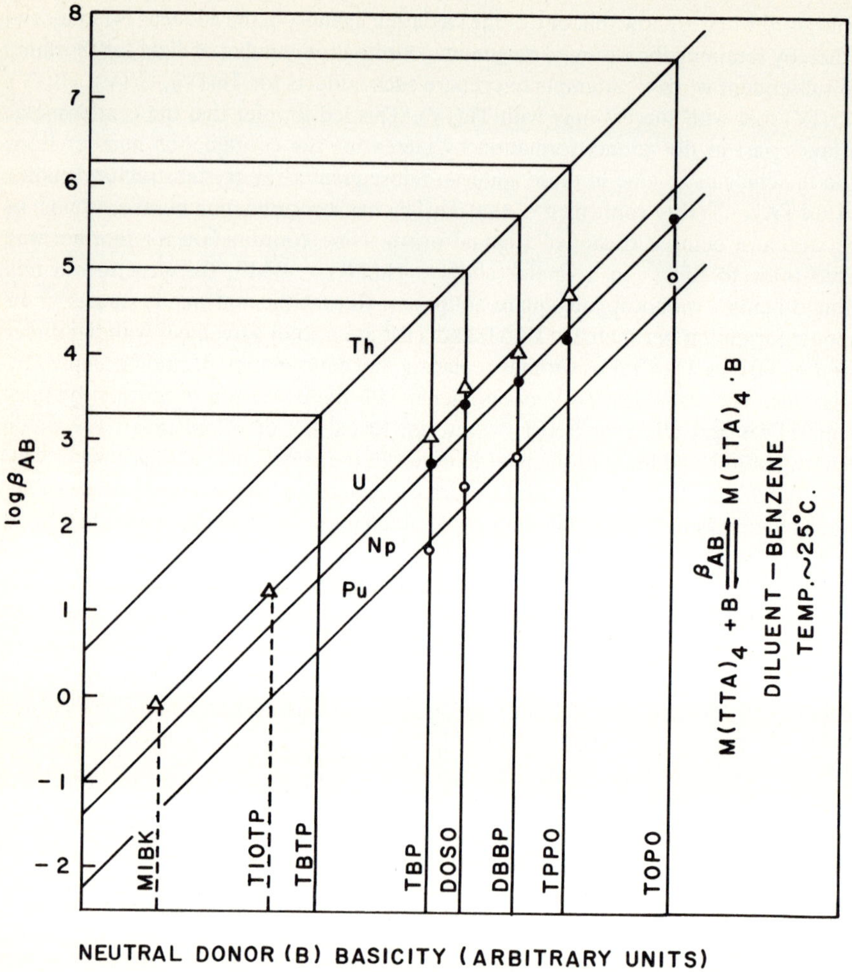

Fig. 6. The relationship between the $M(TTA)_4$ – neutral donor adduct formation constants with the neutral donor basicities. (Reprinted from the report BARC-1108 (1981) by courtesy of the authors)

by a betadiketone-neutral donor combination is probably different from that extracted, under similar conditions, in a non-polar diluent[125]. However, the diluent effect persists in the extraction systems even where the composition of the extracted species was established to the same in all the diluents[10, 83–87, 103, 123–125, 129]. In earlier studies with betadiketone-neutral donor combinations, the decrease in synergic enhancement with increasing polarity of the diluents was attributed[10, 213, 241], among other possibilities, to the destruction of the more extractable anhydrous adduct $MA_n \cdot yB$ by water, probably forming poorly extracted $MA_n \cdot yH_2O$, as more polar diluents extract more water. This reasoning was, however, not favoured[48] as the activity of water in any diluent, that is in equilibrium with the same aqueous solution, is essentially constant. Moreover, the diluents have similar effect on the organic phase adduct formation equilibria (Eq. 17) which were studied using the solutions which were more or less anhydrous and were not

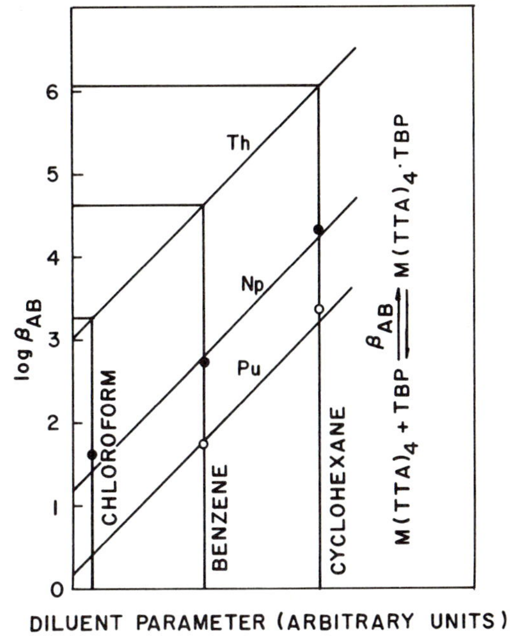

Fig. 7. The relationship between the M(TTA)$_4$-TBP adduct formation constants with the diluents. (Reprinted from the report BARC-1108 (1981) by courtesy of the authors)

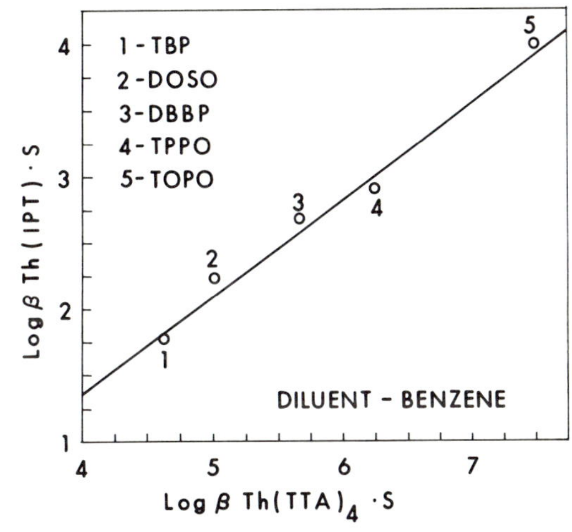

Fig. 8. The relationship between the formations constants of the neutral donor adducts of M(TTA)$_4$ and M(IPT)$_4$. (Reprinted from Ref. 111 by courtesy of the authors)

in equilibrium with any aqueous phase[120]. More work is necessary to understand the role of water, if any, in suppressing synergism. The ability of a diluent to extract water probably indicates its ability of interaction with various species present in it, thereby affecting their activities. Successful application, in some systems[83], of regular solution theory to explain diluent effect tends to support this view.

In the extraction of a metal ion by chelating acids, as well as organophosphorus acids, addition of increasing amounts of a neutral donor results in an initial increase in extraction (synergism), followed by a fall in the enhancement (destruction of synergism) and subsequently antagonism, at higher neutral donor concentrations. This behaviour was attributed, in organophosphorus acid-neutral donor systems, to the interactions between them. With betadiketone-neutral donor systems, however, the extracted water was suggested[241] to be playing a major role in the destruction of the anhydrous adduct, $MA_n \cdot yB$. The similarity in the effect of different diluents and higher concentrations of neutral donor for the same diluent, was demonstrated[241]. The absorption spectral work with at least the adducts of U(IV)[30, 119, 120] and Pu(IV)[52] did not provide any support for the destruction of the adducts in presence of excess water saturated neutral donor. Thus the destruction of synergism at higher neutral donor concentrations appears to be identical to the diluent effect, since at higher neutral donor concentrations an appreciable fraction of the diluent is replaced by the neutral donor which makes the resulting system more polar.

Both these effects must, therefore, be traced to mutual interactions in the organic phase, of the different components present in it, affecting their activities, rather than any destruction of the anhydrous synergic adduct by water.

E. Applications

For the commercial recovery of uranium from wet process phosphoric acid, Oak Ridge National Laboratory, USA, has developed a process[242] using mixtures of HDEHP and TOPO. Spain also has successfully demonstrated a similar process[243]. As the high cost and unavailability of TOPO in commercial quantities are considered as disadvantages of the above process[244], it was suggested that a mixture of HDOPP and TBP is a suitable and much cheaper alternative[245–247].

Mixtures of betadiketones and neutral organophosphorus reagents were employed[118] in the separation of U(VI) from a bulk of rare earths. The antagonistic effect of carboxylic acids, in the high aqueous acid region, on the extraction of U(VI) by TBP was utilized[37] in stripping uranium from TBP. The antagonostic effect of neutral donors on the extraction of thorium and Np(IV) by H_2MEHP was utilized[176] in stripping them from H_2MEHP.

HB_z-TOPO mixtures were used[248] in the separation of thorium and uranium from rock samples. U(VI) in the range 20–50 μg was separated[13] in a single extraction using salicylic acid-TOPO mixtures from 100-fold excess of metals such as rare earths, thorium, zirconium and iron which interfere with the photometric determination of U(VI) by arsenazo III. Methods involving extraction and subsequent spectrophotometric determination were developed for U(VI) using the combinations HDBM-TOPO[249], HAA-pyridine[250], HB_z-rhodamine B[251–253], pyrocatechol-aniline[254] and 2-naphthohydromamic acid-gossipol[255] and for Pu(VI) by HTTA-TBP[99] and for Th(IV) by HDBHQ-rhodamine[199]. HDEHP-TBP mixtures were also used[256] for the extraction of U(VI) from hydrochloric or sulphuric acids for separating it from different anions and cations which interfere in its determination by a fluorescence method.

Methods were developed for the separation of Am and Cm using HPMBP-TOPO[143], for the separation of the Am-Cm pair from rare earths using HTTA-TBP[135], for the separation of Am and Np using HPMBP-TBP[143] and for the separation of Ac from Ra using HTTA-TOPO[257].

The volatility of many neutral donor adducts of fluorine containing betadiketonates have potential application in gas-chromatography[258]. The combination of the specificity of solvent extraction and the resolution capabilities of gas chromatography should considerably broaden their application in the analysis of cationic mixtures by gas chromatography[118]. Some results with such adducts of Th(IV), U(IV)[118], Am(III)[259] and Am(III) and Cm(III)[260], were reported.

The sulphoxide adducts of uranylbetadiketonates were shown[261] to be good nmr shift reagents, which improve the nmr observing frequency by a few orders of magnitude.

The metal-ligand stability constants in aqueous solutions for different metal ions, including Am(III) were determined[262] using some neutral donors in combination with HTTA as the extraction agent for the uncomplexed metal ion.

F. Conclusions

Numerous data on the extraction of actinide ions by mixtures of different extractants have clearly established that synergism is due to the formation of more extractable aquophobic complexes while the destruction of synergism or antagonism is due to the medium effect. The use of diluents such as saturated hydrocarbons, which are more close to "inert diluents" usually gave higher synergism, under otherwise identical conditions. Chelating acids such as betadiketones, which extract the metal ions as their monomeric neutral chelates in different diluents, resulted in a better synergism as compared to other organic acids such as organophosphorus or carboxylic acids. Antagonism is also less when chelating acids are used in combination with neutral extractants apparently due to lesser interactions between the two. Whereever antagonism is observed it is more with more polar neutral extractants such as TOPO, which is in confirmity with its higher interacting ability due to its higher basicity.

A cursory look at various data summarized in the present review reveals that bulk of the work has been devoted to U(VI) and Th(IV) extractions using the combinations of betadiketones and neutral extractants. This is presumably due to the cease of availability of these actinides, their exceptional stability of oxidation states resulting in simplicity in working with them, and large synergic enhancement with such extractant combinations. The misconception that tetravalent actinide ions, forming eight-coordinate chelates with betadiketones would not show synergic enhancement, was probably responsible for the earlier lack of such studies with tetravalent actinides. It is now well-realized that these ions too exhibit synergism and such work is being reported recently, at least using chelating acid-neutral donor combinations. From the trends given in this report it appears likely that even Hf(IV) and Zr(IV) should exhibit synergism with such combinations.

The equilibria involved in the extraction of the actinide ions by neutral donor combinations with chelating extractants such as betadiketones is better understood compared to those when other combinations are used. This is primarily due to the limited number of actinide species extracted and to the monomeric nature of the extractant and the extracting species, thereby making the quantitative interpretation of the extraction data relatively easy. However, it is seen from the data reported here, that even with such extractant combinations the interpretation of the data with tri- and tetravalent actinide ions is quite an involved job. With these ions, the use of aqueous media, such as nitrate or thiocyanate and to a lesser extent chloride or perchlorate, can result in the extraction of many types of species, determining the composition of which is a formidable job. Though many mixed adducts, involving the aqueous anion participation, were established to get extracted, much more work is needed to establish their compositions more precisely.

In synergic systems when the actinide species extracted are limited to the neutral donor adducts of the metal chelates, for a given actinide ion and chelating acid, the adduct stabilities were found to increase with increasing basicity of the neutral donor, and, for a particular donor they, more or less, increase with increasing acidity of the metal ion, as well as of the chelating acid. It has been concluded with reasonable confidence, that, in all these adducts, the monodentate bonding of the chelating ligand, which was strongly believed at one time to satisfy the assumed even-coordination number of the actinide ion, does not exist. Seven-coordination for uranium in uranyl adducts and nine-coordination for the tetravalent ions Th(IV), U(IV), Np(IV) and Pu(IV) seem to be the general rule, in these adducts. The reported adducts of metal chelates with neutral donors with stoichiometries different from 1:1 composition for U(VI)[5, 9, 10, 213, 214], Th(IV)[48] and U(IV)[118] are doubtful and probably non existing.

With many other combinations, the composition of the species extracted are mostly uncertain and the equilibria involved are often very complex involving polymeric species, many times. However many of these systems hold promise of offering useful separation schemes since in some of these systems using carboxylic acids, for example, the extraction and back extraction (stripping) of the metal ions can be achieved in the same system simply by controlling the pH of the aqueous medium.

The knowledge of enthalpy and entropy data on the equilibria involved in synergism are useful in understanding the prevailing mechanism out of various alternatives proposed to explain synergism. Such data available on the systems presented here are very little and even that too was obtained by measurements at different temperatures. If such data is generated for more systems, preferably by calorimetry it would be of much use in understanding the type of species involved in synergism.

In most chelate complexes of lanthanides and actinides the metal-oxygen bond is expected to be predominantly electrostatic[263]. Spectroscopic evidence, however, suggests[264] that some degree of covalency is present in these complexes and that the covalency expected[265] in actinide bonding is larger than in lanthanide bonding owing to less shielding of 5f orbitals than 4f orbitals. With the exception of Th(IV) and U(VI) most of the actinide ions have partially filled f-configuration and the magnetic measurements on their synergic complexes would be useful in a better understanding of their bonding. Spectroscopic studies of these complexes also would be useful in this direction. Unfortunately, hardly any work has been devoted to such studies.

The species involved in the synergic extraction of actinide ions present interesting coordination compounds for studying their structures and unterstanding the bonding. However, the information available on the structures of such complexes is very limited. A beginning, however, has been made recently to investigate synergic complexes of actinides[226, 237] and of other metal ions[266–269]. As pointed out earlier, the coordination number of the actinide atom in synergic complexes can be as high as at least nine. It is known that relatively few complexes are known with such high coordination numbers and synergic complexes of the actinide ions provide a number of such high coordination complexes. It is known that two factors favour the formation of high coordination complexes. They are (i) the size of the metal cation which has to be significantly large to provide at least nine coordiation sites, without undue crowding of the ligands, and (ii) the metal has to be in a high formal oxidation state. Actinide ions meet both these requirements as is evident from a number of high coordination number complexes reported for them[270]. Except for $Th(TTA)_4 \cdot TOPO$, no structural work has been reported for any tetravalent ion complexes involved in synergism. The situation is almost similar for hexavalent actinides, where only a few U(VI) complexes, with seven-coordination, involved in synergism, have been subjected to structural studies. As can be seen from this review it is clear that some mixed adducts of the type $ML_xA_{(n-x)} \cdot aB$ involving the participation of the aqueous anion in the complex are formed during the extraction of tri- and tetravalent actinides, under certain conditions. Very few studies even by solvent extraction, have been devoted to this area. As not even a single complex of this type has been isolated so far their structural aspects remain unstudied. In such complexes of tetravalent actinides such as the $Pu(TTA)_3(NO_3) \cdot 2TOPO$, a coordination number of even ten may be exhibited. Thus, the isolation and investigation of such complexes would be highly rewarding. While this type of mixed adducts are common in the synergic extraction of tri- and tetravalent actinides, no such adducts appear to be formed in the extraction of U(VI). As a matter of fact, no complex of the type $UO_2AA' \cdot B$, where A and A′ are two different bidentate ligands and B is a neutral donor appears to have been reported. The investigation of such complexes would be of great interest to know why a complex of the type $UO_2(NO_3)(TTA) \cdot TBP$ is not formed at all.

It is to be hoped that with improved three-dimensional x-ray techniques and with ever increasing interest in the coordination chemistry of actinides, more attention from a coordination chemist will be drawn to these fascinating complexes.

G. Abbreviations

I. Neutral Extractants

TOPO	tri-n-octyl phosphine oxide
TBPO	tri-n-butyl phosphine oxide
TEHPO	tri-2, ethylhexyl phosphine oxide
TPPO	tri-phenyl phosphine oxide
BDBP	n-butyl, di(n-butyl) phosphinate
DBBP	di(n-butyl), n-butyl phosphonate
DAMP	di(isopantyl), methyl phosphonate
DEEP	di(ethyl), ethyl phosphonate
TMP	trimethylphosphate
TBP	tri-n-butyl phosphate
TEP	tri ethyl phosphate
TPrP	tri propyl phosphate
DBDECMP	dibutyl-N, N-diethylcarbamoyl methylene phosponate
TPP	tri phenyl phosphate
TBTP	tributyl thio phosphate
TIOTP	tri isooctyl thio phosphate
TLAO	tri lauryl amine oxide
PO	Pyridine-N-oxide
2MPO	2-methyl, pyridine-N-oxide

3 MPO	3-methyl, pyridine-N-oxide
4 MPO	4-methyl, pyridine-N-oxide
4 MOPO	4-methoxy, pyridine-N-oxide
4 CPO	4-chloro pyridine-N-oxide
4 NPO	4-nitro pyridine-N-oxide
2,6-DMPO	2,6-dimethyl, pyridine-N-oxide
DMSO	dimethyl sulphoxide
D-n-PSO	di(n-pentyl) sulphoxide
DHSO	dihexyl sulphoxide
DOSO	dioctyl sulphoxide
DISO	di(isooctyl) sulphoxide
DDSO	didecyl sulphoxide
DBSO	dibenzyl sulphoxide
DPSO	diphenyl sulphoxide
MIBK	methyl isobutyl ketone
DMF	dimethyl formamide
DBE	di-n-butyl ether
DIBE	di-(isobutyl) ether
DBS	dibutyl thio ether
$DBSO_2$	dibutyl sulfone
DPE	diphenyl ether
DPS	diphenyl thio ether
DPSe	diphenyl seleno ether
n-BA	n-butylalcohol
N Bu	nitrobutane
N Be	nitrobenzene
DEA	N,N-diethylacetamide
DIPY	di-pyridyl
PHEN	O-phenanthroline

II. *Acidic Extractants*

HAA	acetyl acetone
ClAA	3-chloro, pentane 2,4-dione
HBA	benzoyl acetone
HTA	thenoyl acetone
HTFA	trifluoro acetyl acetone
HHFA	hexafluoro acetyl acetone
HBTA	benzoyl trifluoro acetone
HFTA	furoyl trifluoro actone
HTTA	thenoyl trifluoro acetone
HTTMA	thenoyl trifluoro (methyl) acetone
HDBM	dibenzoyl methane
HTR	Tropolone
HIPT	β-isopropyl tropolone
HO_x	oxine (8-hydroxy quinoline)
HDBHQ	5,7-dibromo, 8-hydroxy-quinoline
HCF	Cupferron
HBPHA	N-benzoyl, N-phenyl, hydroxylamine
HPMBP	1-phenyl, 3-methyl, 4-benzoyl, pyrazolone-5
HDEHP	di(2-ethylhexyl) phosphoric acid
HDBP	di(butyl) phosphoric acid
HDAP	di(isopentyl) phosphoric acid
HAMP	isopentyl methyl phosphoric acid
HDOPP	di(4-octyl phenyl) phosphoric acid
DIDP	di(isodecyl) phosphoric acid
H_2MEHP	mono, 2-ethylhexyl, phosphoric acid
H_2MBP	mono, butyl, phosphoric acid
HB_z	benzoic acid
HDNNS	di, nonyl, naphthalene sulphonic acid
HDMPP	3,5-dibutyl, 1-methyl, 4-propylpyrazole

III. *Basic Extractants*

ALIQ	aliquat-336, a quaternary amine
Q	capriquat (or)
TOMA	Tri-n-octyl methyl ammonium, a quaternary amine
TOA	tri-n-octyl amine
TLA	tri lauryl amine
MNDDA	N-methyl didecyl amine

Acknowledgements. The authors sincerely thank Dr. M. V. Ramaniah, Director, Radiological Group and Dr. P. R. Natarajan, Head, Radiochemistry Division for their keen interest in this work.

H. References

1. Baes, C. F., Jr.: Nucl. Sci. Engg. *16*, 405 (1963)
2. Blake, C. A. et al.: Proc. 2nd Int. Conf. PUAE, Geneva 1958, *28*, 289 (1959)
3. Cunninghame, J. G., Scargill, P., Willis, H. M.: AERE C/M 215 (1954)
4. Irving, H., Edgington, D. N.: Proc. Chem. Soc. 360 (1959)

5. Irving, H., Edgington, D. N.: J. Inorg. Nucl. Chem. *15*, 158 (1960)
6. Irving, H., Edgington, D. N.: ibid. *20*, 314 (1961)
7. Irving, H., Edgington, D. N.: ibid. *20*, 321 (1961)
8. Irving, H., Edgington, D. N.: ibid. *21*, 169 (1961)
9. Healy, T. V.: ibid. *19*, 314 (1961)
10. Healy, T. V.: ibid. *19*, 328 (1961)
11. Healy, T. V., Peppard, D. F., Mason, G. W.: ibid. *24*, 1429 (1962)
12. Irving, H. M. N. H.: Solvent Extraction Chemistry (Dyrssen, D., Liljenzin, O., Rydberg, J., Eds.) North-Holland, Amsterdam 1967, p. 91
13. Mareva, St., Jordanov, N., Konstantinova, M.: Anal. Chim. Acta *59*, 319 (1972)
14. Sekine, T. and Hasegawa, Y.: Solvent Extraction Chemistry, Marcel Dekker Inc., New York 1969, p. 217
15. Healy, T. V.: Solvent Extraction Research (Kertes, A. S., Marcus, Y., Eds.) Wiley, New York 1969, p. 815
16. Marcus, Y. and Kertes, A. S.: Ion Exchange and Solvent Extraction of Metal Complexes, Wiley, New York 1969, p. 815
17. Duyckaerts, G., Desreux, J. F.: Proc. ISEC 77, CIM Sp. vol. 21, p. 1 (1978)
18. Healy, T. V.: Gmelin Handbook of Inorganc Chemistry, part D_2 (1975)
19. Newman, L.: J. Inorg. Nucl. Chem. *25*, 304 (1963)
20. Taube, M., Siekierski, S.: Nukleonica *6*, 489 (1961)
21. Dyrssen, D.: Proc. 8 ICCC, Vienna 1964, p. 362 (1964)
22. Gal, I. J., Nikolic, R. M.: J. Inorg. Nucl. Chem. *28*, 563 (1966)
23. Sekine, T., Hasegawa, Y.: Solvent Extraction Chemistry, Marcel Dekker Inc., New York 1969, p. 211
24. Patil, S. K., Ramakrishna, V. V., Kartha, P. K. S., Gudi, N. M.: Sep. Sci. Technol. *15*, 1459 (1980)
25. Patil, S. K., Ramakrishna, V. V., Haraprakas, B.: J. Inorg. Nucl. Chem. *43*, 1377 (1981)
26. Healy, T. V., Ferraro, J. R.: ibid. *24*, 1449 (1962)
27. Butts, W. C., Banks, C. V.: Anal. Chem. *42*, 133 (1970)
28. Graddon, D. P., Watton, E. C.: J. Inorg. Nucl. Chem. *21*, 49 (1961)
29. Dyrssen, D., Petkovic, D.: Acta. Chem. Scand. *19*, 653 (1965)
30. Ramakrishna, V. V., Patil, S. K., Ramanujam, A.: J. Inorg. Nucl. Chem. *42*, 1009 (1980)
31. Dyrssen, D., Falk, K. E., Ivanova, E. K.: Acta Chem. Scand. *26*, 3865 (1972)
32. Rao, G. N., Li, N. C.: J. Inorg. Nucl. Chem. *28*, 2931 (1966)
33. Desreux, J. F., Massaux, J., Duyckaerts, G.: ibid. *40*, 1159 (1978)
34. Bullock, J. I., Sweatman, A. J.: J. Inorg. Chim. Acta *64*, LI (1982)
35. Nakamura, K.: J. Inorg. Nucl. Chem. *31*, 455 (1969)
36. Sajun, M. S.: Radiochemical Study of Actinide Elements, Ph.D. Thesis, Bombay Univ. 1981
37. Kassierer, E. F., Kertes, A. S.: J. Inorg. Nucl. Chem. *34*, 3209 (1972)
38. Drago, R. S., Purcell, K. F.: Prog. Inorg. Chem. *6*, 271 (1964)
39. Wakabayashi, T., Oki, S., Omiri, T., Suzuki, N.: J. Inorg. Nucl. Chem. *26*, 2255 (1964)
40. Akiba, K., Suzuki, N., Asano, H., Kanno, T.: J. Radioanal. Chem. *7*, 203 (1971)
41. Tournier, R. A., Davis, M. W., Jr.: Sep. Sci. *7*, 159 (1972)
42. Moshinura, Y., Suzuki, N.: Anal. Chim. Acta *85*, 383 (1976)
43. Akiba, K., Wada, M., Kanno, T.: Res. Reg. Lab. Nucl. Sci. Tohoku Univ. *10*, 268 (1977)
44. Akiba, K., Wada, M., Kanno, T.: J. Inorg. Nucl. Chem. *42*, 261 (1980)
45. Dyrssen, D., Petkovic, DJ.: ibid. *27*, 1381 (1965)
46. Nakamura, K.: ibid. *32*, 2265 (1970)
47. Petkovic, DJ. M., Kezele, B. A.: Proc. ISEC 71, Soc. Chem. Ind., London *2*, 1137 (1971)
48. Sekine, T., Dyrssen, D.: J. Inorg. Nucl. Chem. *29*, 1457 (1967)
49. Akiba, K., Suzuki, N., Kanno, T.: Bull. Chem. Soc. Japan *42*, 2537 (1969)
50. Pukanic, G., Li, N. C., Brey, W. S., Jr., Savitsky, G. B.: J. Phys. Chem. *70*, 2899 (1966)
51. Cox, E. C., Jr., Davis, M. W., Jr.: Sep. Sci. *8*, 205 (1973)
52. Ramakrishna, V. V., Patil, S.K., Haraprakas, B.: Sep. Sci. Technol. *14*, 571 (1979)
53. Mathur, J. N., Khopkar, P. K.: Talanta *29*, 633 (1982)
54. Newman, L., Klotz, P.: J. Phys. Chem. *65*, 796 (1961)
55. Dyrssen, D., Hennicks, M.: Acta Chem. Scand. *15*, 47 (1961)

56. Newman, L., Klotz, P.: J. Phys. Chem. *66*, 2262 (1962)
57. Inoue, Y., Tochiyama, O., Oda, J.: J. Inorg. Nucl. Chem. *41*, 1375 (1979)
58. LeRoux, H. J., Fouche, K. F.: ibid. *34*, 747 (1972)
59. Konstantinova, M., Mareva, St., Jordanov, N.: Anal. Chim. Acta *90*, 295 (1977)
60. Madic, C., Koehly, G.: Nucl. Technol. *41*, 323 (1978)
61. Chmutova, M. K., Myasoedov, B. F., Spivakov, B. Y., Koechetkova, N. E., Zolotov, Y. A.: J. Inorg. Nucl. Chem. *35*, 1317 (1973)
62. Marcus, Y., Kertes, A. S.: Ion Exchange and Solvent Extraction of Metal Complexes, Wiley, New York 1969, p. 836
63. D. H. Liem: On the Solvent Extraction of Metal Ions by Dialkyl Phosphoric Acid and Some Non-Acidic Complexing Agents, Stockholm 1971
64. Rydberg, J.: Ark. Kimi *8*, 113 (1955)
65. Krishen, A., Freiser, H.: Anal. Chem. *29*, 288 (1957)
66. Lundquist, R.: Proc. ISEC 1980, paper 80-51
67. Stary, J., Collect. Czech. Chem. Commun. *25*, 890 (1960)
68. Mouca, V., Stary, J.: ibid. *26*, 763 (1961)
69. Schmeitzer, G. K., Moltern, J. L.: Anal. Chim. Acta *26*, 120 (1962)
70. Orr, W. C.: UCRL-196 (1948)
71. Larsen, E. M., Terry, G., Leddy, J.: J. Am. Chem. Soc. *75*, 5107 (1953)
72. Day, R. A., Jr., Powers, R. M.: ibid. *76*, 3895 (1954)
73. King, E. L.: TID-5290, paper 34, 1958, p. 269
74. Peterson, S.: J. Inorg. Nucl. Chem. *14*, 126 (1960)
75. Dyrssen, D.: Trans. Roy. Inst. Technol. No. 188, Stockholm (1962)
76. Godbole, A. G., Swarup, R., Ramakrishna, V. V., Patil, S. K.: Radiochem. Radioanal. Lett. *56*, 195 (1983)
77. Stary, J.: Anal. Chim. Acta *28*, 132 (1963)
78. Stary, J., Aladky, E.: ibid. *28*, 227 (1963)
79. Oki, S.: ibid. *44*, 315 (1969)
80. Thakur, N. V.: Proc. DAE Nucl. Chem. Radiochem. Symp. Waltair, India 1980, p. 383
81. Bazdar, K., Goldberg, D. E., Newman, L.: J. Inorg. Nucl. Chem. *29*, 1511 (1967)
82. Healy, T. V.: ibid. *31*, 499 (1969)
83. Akiba, K., Suzuki, N.: Bull. Chem. Soc. Japan *44*, 1043 (1971)
84. Akiba, K., Hashimoto, H., Kanno, T.: Res. Rep. Lab. Nucl. Sci. Tohoku Univ. *4*, 126 (1971)
85. Akiba, K., Hashimoto, H., Kanno, T.: ibid. *5*, 167 (1972)
86. Akiba, K.: J. Inorg. Nucl. Chem. *35*, 3323 (1973)
87. Akiba, K.: J. Radioanal. Chem. *36*, 153 (1977)
88. Milich, N., Petrukhin, O. M., Zolotov, Yu. A.: Zh. Neorg. Khim. *9*, 2664 (1964)
89. Subramanian, M. S., Viswanatha, A.: J. Inorg. Nucl. Chem. *31*, 2575 (1969)
90. Subramanian, M. S., Pai, S. A.: ibid. *32*, 3677 (1970)
91. Subramanian, M. S., Pai, S. A.: Aust. J. Chem. *26*, 77 (1973)
92. Manchanda, V. K., Shukla, J. P., Subramanian, M. S.: J. Inorg. Nucl. Chem. *36*, 2595 (1974)
93. Manchanda, V. K., Shukla, J. P., Subramanian, M. S.: J. Radioanal. Chem. *29*, 69 (1976)
94. Puzic, R. G., Maksimouic, Z. B.: Proc. ISEC 1971, Soc. Chem. Ind. London *2*, 1180 (1971)
95. Patil, S. K., Bhandiwad, V., Kusumakumari, M., Swarup, R.: J. Inorg. Nucl. Chem. *43*, 1647 (1981)
96. Ramakrishna, V. V., Swarup, R., Patil, S. K.: Proc. Indian Acad. Sci. (Chem. Sci.) *91*, 499 (1982)
97. Bakos, L. Szabo, E., Andras, L., Iyer, N.: Proc. Symp. Coordin. Chem. Tihany, Hungary, 1964 (Beck, M. T., ed.), Akademiai Kaido, Budapest 1965, p. 241
98. Patil, S. K., Godbole, A. G., Swarup, R.: J. Radioanal. Chem. *60*, 31 (1980)
99. Shukla, J. P., Manchanda, V. K., Subramanian, M. S.: ibid. *29*, 61 (1976).
100. Genov, L., Kasalov, G.: Monatsch. Chem. *100*, 594 (1969)
101. Awwal, M. A., Shamsuddin, A. K.: Nucleus (Karachi) *7*, 190 (1970)
102. Patil, S. K., Ramakrishna, V. V., Sajun, M. S., Thiagarajan, R.: Sep. Sci. Technol. *14*, 859 (1979)
103. Patil, S. K., Ramakrishna, V. V., Sajun, M. S., Thiagarajan, T.: J. Radioanal. Chem. *54*, 159 (1979)

104. Patil, S. K., Ramakrishna, V. V., Sajun, M. S.: Radiochem. Radioanal. Lett. *44*, 239 (1980)
105. Sajun, M. S., Ramakrishna, V. V. , Patil, S. K.: J. Radioanal. Chem. *63*, 57 (1981)
106. Taketatsu, T., Toriumi, N.: J. Inorg. Nucl. Chem. *31*, 2235 (1969)
107. Taketatsu, T., Ohkura, N.: Bull. Chem. Soc. Japan *44*, 2430 (1971)
108. Finston, H. L., Gnizi, E., in: Solvent Extraction Research (Kertes, A. S., Marcus, Y., Eds.), Wiley-Interscience, New York 1969, p. 333
109. Yeh, S., Chu, P.: J. Chin. Chem. Soc. (Taipei) *10*, 1 (1963)
110. Kassierer, E. F., Kertes, A. S.: J. Inorg. Nucl. Chem. *34*, 778 (1972)
111. Mapara, P. M. et al.: J. Radioanal. Chem. (In Press)
112. Auilana, V. N., Driomko, V. M.: Symp. Chem. Ext. Proc. Izd. Nauka, Moscow 1972
113. Abraham, J., Corsini, A.: Anal. Chem. *42*, 1528 (1970)
114. Wallace, W. J., Gulub, A. M., Dabeka, R. W.: XV ICCC, Moscow 1973, p. 292
115. Dyrssen, D., Dahlberg, V.: Acta Chem. Scand. *7*, 1186 (1953)
116. Hagiwara, Z.: Tohoku Univ. Tech. Rep. *18*, 16 (1953)
117. Bok, L. D. C., Wessels, G. F. S., Leipoldt, J. G.: Z. Anorg. Allg. Chem. *404*, 76 (1974)
118. Mitchell, J. W.: Ph.D. Thesis, Iowa State Univ. IS-T-353 (1970)
119. Patil, S. K., Ramakrishna, V. V., Ramanujam, A.: Proc. Indian Acad. Sci. *88 A*, 303 (1979)
120. Patil, S. K., Ramakrishna, V. V.: J. Radioanal. Chem. *54*, 123 (1979)
121. Ramanujam, A., Gudi, N. M., Nadkarni, M. N., Patil, S. K., Ramakrishna, V. V.: Proc. Indian Acad. Sci. (Chem. Sci.) *91*, 347 (1982)
122. Bullock, J. I., Sweatman, A. J.: Inorg. Chim. Acta *64*, L_1 (1982)
123. Ramanujam, A., Ramakrishna, V. V., Patil, S. K.: Sep. Sci. Technol. *14*, 13 (1979)
124. Patil, S. K., Ramakrishna, V. V., Gudi, N. M.: Radiochem. Radioanal. Lett. *38*, 363 (1979)
125. Patil, S. K., Ramakrishna, V. V., Gudi, N. M.: Sep. Sci. Technol. *15*, 1513 (1980)
126. Patil, S. K., Ramakrishna, V. V., Kartha, P. K. S., Gudi, N. M.: J. Radioanal. Chem. *59*, 327 (1980)
127. Sajun, M. S., Ramakrishna, V. V., Patil, S. K.: Radiochem. Radioanal. Lett. *47*, 67 (1981)
128. Sajun, M. S., Ramakrishna, V. V., Patil, S. K.: Proc. Indian Acad. Sci. (Chem. Sci.) *90*, 323 (1981)
129. Patil, S. K., Ramakrishna, V. V., Haraprakas, B.: Sep. Sci. Technol. *15*, 131 (1980)
130. Ramakrishna, V. V., Mapara, P. M., Patil, S. K.: J. Radioanal. Chem. *78*, 65 (1983)
131. Newman, L., Klotz, P.: Inorg. Chem. *5*, 461 (1966)
132. Sekine, T., Dyrssen, D.: J. Inorg. Nucl. Chem. *29*, 1481 (1967)
133. Schreck, H.: KFK-672 (1967)
134. Feinauer, D.: KFK-1298 (1970)
135. Davis, M. W., Jr., Holt, J. W., Tournier, R. A.: Proc. ISEC 1971, Soc. Chem. Ind. London, Paper 69, 1971, p. 974
136. Ramakrishna, V. V., Patil, S. K., Krishnareddy, L., Reddy, A. S.: J. Radioanal. Chem. *47*, 57 (1978)
137. Ramakrishna, V. V., Patil, S. K., Krishnareddy, L., Reddy, A. S.: J. Inorg. Nucl. Chem. *41*, 113 (1979).
138. Wada, M., Akiba, K., Kanno, T.: Res. Inst. Min. Dress. Met. Rep. No. 784 (1979)
139. Khopkar, P. K., Mathur, J. N.: J. Radioanal. Chem. *60*, 131 (1980)
140. Khopkar, P. K., Mathur, J. N.: Sep. Sci. Technol. *16*, 957 (1981)
141. Ramanujam, A., et al.: DAE Nucl. Chem. Radiochem. Symp., Banaras, India 1981, Paper CT-11
142. Healy, T. V.: Nucl. Sci. Eng. *16*, 413 (1963)
143. Myasoedov, B. F., Chmutova, M. K., Lebedev, I. A.: Proc. ISEC 1971, Soc. Chem. Ind. London 1971, p. 815
144. Bacher, W., Keller, C.: J. Inorg. Nucl. Chem. *35*, 2945 (1973)
145. Khopkar, P. K., Mathur, J. N.: Sep. Sci. Technol. *17*, 985 (1982)
146. Blake, C. A., Horner, D. E., Schmidt, I. M., ORNL-2259 (1959)
147. Dyrssen, D., Kuca, L.: Acta Chem. Scand. *14*, 1945 (1960)
148. Kennedy, J., Burford, F. A., Sammes, P. G.: J. Inorg. Nucl. Chem. *14,* 114 (1960)
149. Shevchenko, V. B., Smelov, V. S., Strakhova, A. V., in: Ekstraktziya, vol. 2. Gosatomizdat, Moscow 1960, p. 179
150. Kennedy, J., Deane, A. M.: J. Inorg. Nucl. Chem. *19,* 142 (1961)

151. Shevchenko, V. B., Smelov, V. S., Strakhova, A. V.: Russ. J. Inorg. Chem. *7*, 894 (1962)
152. Zangen, M.: J. Inorg. Nucl. Chem. *25*, 581 (1963)
153. Ihle, H., Michael, M., Murrenhoff, A.: ibid. *25*, 734 (1963)
154. Hahn, H. T., Vanderwall, E. M.: ibid. *26*, 191 (1964)
155. Sato, T.: ibid. *26*, 311 (1964)
156. Liem, D. H., Dyrssen, D.: Acta Chem. Scand. *20*, 272 (1966)
157. Liem, D. H.: ibid. *22*, 773 (1968)
158. Bykhovtsov, V. L., Melnikova, G. N.: Radiokhimiya *11*, 619 (1969)
159. Bykhovtsov, V. L.: ibid. *12*, 412 (1970)
160. Bykhovtsov, V. L., Zimina, T. Ya.: ibid. *12*, 686 (1970)
161. Vavilov, N. G.: ibid. *15*, 752 (1973)
162. Marcus, Y., Kolarik, Z.: Inorg. Nucl. Chem. Lett. *10*, 275 (1974)
163. Torgov, V. G., Us, T. V., Mikhailov, V. A., Nikolaev, A. V.: Dokl. Akad. Nauk. SSSR *227*, 635 (1976)
164. Torgov, V. G., Us, T. V., Glinskaya, A. N., Mikhailov, V. A.: Ser. Khim. Nauk *4*, 63 (1977)
165. Bunus, F. T., Domocos, V. C., Dumitriescu, P.: J. Inorg. Nucl. Chem. *40*, 117 (1978)
166. Martin, J. L. M., Leroy, M. J. F.: J. Chem. Res. (M), 1113 (1978)
167. Nagle, R. A., Murty, T. K. S.: Sep. Sci. Technol. *14*, 499 (1979)
168. Zangen, M.: J. Inorg. Nucl. Chem. *28*, 1693 (1966)
169. Mason, G. W., McCarty, S., Peppard, D. F.: ibid. *24*, 967 (1962)
170. Deptula, C.: Nucleonica *12*, 211 (1967)
171. Deptula, C., Minc, S.: J. Inorg. Nucl. Chem. *29*, 159 (1967)
172. Deptula, C.: ibid. *32*, 277 (1970)
173. Zangen, M.: ibid. *25*, 1051 (1963)
174. Bykhovtsov, V. L.: Radiokhimiya *12*, 539 (1970)
175. Lee, T., Ting, G.: Sep. Sci. Technol. *16*, 303 (1981)
176. Peppard, D. F., Mason, G. W., Sironen, R. J.: J. Inorg. Nucl. Chem. *10*, 117 (1959)
177. Tachimori, S., Nakamura, H.: J. Nucl. Sci. Technol. *19*, 326 (1982)
178. Kolarik, Z.: J. Inorg. Nucl. Chem. *34*, 2911 (1972)
179. Tachimori, S.: J. Radioanal. Chem. *49*, 31 (1979)
180. Vandalen, A., Gerritsma, K. W.: Proc. ISEC 1971, Soc. Chem. Ind. London 1971, p. 1096
181. Vandalen, A., Wijkstra, J., Gerritsma, K. W.: J. Inorg. Nucl. Chem. *40*, 875 (1978)
182. Konstantinova, M., Mareva, St., Jordanov, N.: Anal. Chim. Acta *68*, 237 (1974)
183. Madic, C.: Ph.D. Thesis, Paris, CEA-R-4702 (1975)
184. Konstantinova, M.: Anal. Chim. Acta *90*, 185 (1977)
185. Nikitin, Yu. E., Kuvatov, Yu. G., Egutkin, N. L., Murinov, Yu. I.: Radiokhimiya *19*, 642 (1977)
186. Sillen, L. G., Martell, A. E. (eds.): Stability Constants of Metal Ion Complexes, The Chem. Soc. London 1964
187. Madic, C., Koehly, G.: CEA-R-4514 (1973)
188. Madic, C., Koehly, G.: Inorg. Nucl. Chem. Lett. *12*, 153 (1976)
189. Madic, C. et al.: J. Inorg. Nucl. Chem. *38*, 2081 (1976)
190. Vieux, A. S., Ngiefu, K., Rutagengwa, N.: ibid. *40*, 1431 (1978)
191. Ryan, J. L., Keder,W. E., in: Advances in Chemistry Series, No. 71, Lanthanide/Actinide Chemistry, Am. Chem. Soc. 1967, p. 335
192. Patil, S. K., Swarup, R., Bhandiwad, V.: J. Radioanal. Chem. *53*, 115 (1979)
193. Hok-Bernstrom, B.: Acta Chem. Scand. *10*, 174 (1956)
194. Bullwinkel, E. P., Noble, P., Jr.: J. Am. Chem. Soc. *80*, 2955 (1958)
195. Nikolic, R. M., Gal, I. J.: Croat. Chim. Acta *38*, 17 (1966)
196. Takei, S.: Nippon Kagaku Zasshi *87*, 949 (1966)
197. Yoshida, H.: Bull. Chem. Soc. Japan *39*, 1810 (1966)
198. Newman, L., Klotz, P.: J. Phys. Chem. *67*, 205 (1963)
199. Mischenko, V. T., Zavarina, T. V.: Zh. Anal. Chim. *25*, 1533 (1970)
200. Mutterties, E. L.: J. Am. Chem. Soc. *88*, 305 (1966)
201. Mapara, P. M., Negi, S., Ramakrishna, V. V., Patil, S. K.: DAE Symp. Radiochem. Radiation Chem. Poona, India, 1982, Paper CT-8
202. Khopkar, P. K., Mathur, J. N.: J. Inorg. Nucl. Chem. *39*, 2063 (1977)

203. Spivakov, B. Y., Zolotov, Y. A., Myasoedov, B. F., Chmutova, M. K., Koechetkova, N. E.: Zh. Neorg. Khim. *17*, 3334 (1972)
204. Kraac, W., Bac, R.: Symp. Chem. Ext. Met. Org. Solv. Atomizdat, Moscow 1969, p. 195
205. Mohanty, S. R., Reddy, A. S.: J. Inorg. Nucl. Chem. *37*, 1791 (1975)
206. Swarup, R., Godbole, A. G., Patil, S. K.: Radiochem. Radioanal. Lett. *54*, 193 (1982)
207. Alina, A., Haggag, A.: Indian J. Chem. *5*, 579 (1967)
208. Ihle, H., Michael, H., Murrenhoff, A.: EUR-2635 (1966)
209. Kolarik, Z., Puzic, R. G., Maksimovic, Z. B.: J. Inorg. Nucl. Chem. *31*, 2485 (1969)
210. Kraac, W., Bac, R.: Solvent Extraction Chemistry of Metals, Proc. Int. Conf. 1965, p. 267
211. Mohanty, S. R., Reddy, A. S.: J. Inorg. Nucl. Chem. *37*, 1977 (1975)
212. Navratil, J. D., Martella, L. L.: Sep. Sci. Technol. *16*, 1147 (1981)
213. Ferraro, J. R., Healy, T. V.: J. Inorg. Nucl. Chem. *24*, 1463 (1962)
214. Li, N. C., Wang, S. M., Walker, W. R.: ibid. *27*, 2263 (1965)
215. Patil, S. K., Ramakrishna, V. V., Sajun, M. S.: Thermochim. Acta *42*, 281 (1980)
216. Rao, G. N., Li, N. C.: Can. J. Chem. *44*, 2775 (1966)
217. Subramanian, M. S., Manchanda, V. K.: J. Inorg. Nucl. Chem. *33*, 3001 (1971)
218. Subramanian, M. S., Pai, S. A., Manchanda, V. K.: Aust. J. Chem. *26*, 85 (1973)
219. Manchanda, V. K., Subramanian, M. S.: ibid. *27*, 1573 (1974)
220. Manchanda, V. K., Subramanian, M. S.: Acta Chim. Acad. Sci. Hung *111*, 69 (1982)
221. Subramanian, M. S., Khopkar, P. K., Shukla, J. P., Pai, S. A.: J. Inorg. Nucl. Chem. *36*, 3862 (1974)
222. Ferraro, J. R.: ibid. *26*, 225 (1964)
223. Levy, J. H., Waugh, A. B.: J. Chem. Soc. Dalton 1628 (1977)
224. Taylor, J. C., Waugh, A. B.: ibid. 1630 (1977)
225. Taylor, J. C., Waugh, A. B.: ibid. 1636 (1977)
226. Lu, T. H., Lee, T. J., Lee, T. Y., Wong, C.: Inorg. Nucl. Chem. Lett. *13*, 363 (1977)
227. Taylor, J. C., McLaren, A. B.: J. Chem. Soc. Dalton 460 (1979)
228. Fleming, J. E., Lynton, H.: Can. J. Chem. *45*, 1637 (1967)
229. Hall, D., Rae, A. D., Waters, T. N.: Acta Cryst. *22*, 258 (1967)
230. Comyns, A. E., Gatehouse, B. M., Wait, E.: J. Chem. Soc. 4655 (1958)
231. Haigh, J. M., Thornton, D. A.: Inorg. Nucl. Chem. Lett. *6*, 231 (1970)
232. Lenner, M.: Acta Cryst. *B35*, 2396 (1979)
233. Sacconi, L., Paoletti, P.: J. Chem. Soc. 4250 (1961)
234. Mentzen, B. F., Puaux, J. P., Loiseleur, H.: Acta Cryst. B*33*, 1848 (1977)
235. Casellato, U., Vigato, P. A., Vidali, M.: Coordin. Chem. Rev. *26*, 85 (1978)
236. Martell, A. E., Calvin, M.: Chemistry of the Metal Chelate Compounds, Prentice Hall, New York 1952, p. 243
237. Leipoldt, J. G., Wessels, G. F. S., Bok, L. D. C.: J. Inorg. Nucl. Chem. *37*, 2487 (1975)
238. Day, V. W., Hoard, J. L.: J. Am. Chem. Soc. *92*, 3626 (1970)
239. Folcher, G. et al.: Can. J. Chem. *55*, 3559 (1977)
240. Patil, S. K., Ramakrishna, V. V., Thakur, N. V.: XX ICCC, Calcutta, India 1979, Abs. No. I-59, p. 145
241. Healy, T. V., Peppard, D. F., Mason, G. W.: J. Inorg. Nucl. Chem. *24*, 1429 (1962)
242. Hurst, F. J., Ryon, A. D.: Chem. Engg. *84*, 56 (1977)
243. Cordero, G. et al.: IAEA-CN-36/211, Salzburg Conf. May, 1977, vol. 2, p. 377 (1977)
244. Ring, R. J.: At. Energy (Australia) Jan. 1977, p. 12
245. Murthy, T. K. S., Pai, V. N., Nagle, R. A.: IAEA Proc. Series, Vienna, 341 (1971)
246. Dar, K. K. et al.: Proc. 4th Int. Conf. PUAE, UN, New York *8*, 106 (1972)
247. Murthy, T. K. S.: IAEA-R-834 F (1972)
248. Thach, N. D., Mareva, St., Jordanov, N.: Commun. Dept. Chem. *3*, 671 (1970)
249. Martin, M.: CEA-CR-11 (1975)
250. Shinde, V. M.: Chem. Anal. (Warsaw) *21*, 813 (1976)
251. De Silva, F., de Moura, L.: Proc. 2nd Int. Conf. PUAE *28*, 480 (1958)
252. Mocken, H. H. Ph., Neste, V.: Anal. Chim. Acta *37*, 480 (1967)
253. Dubey, S. C., Murthy, T. K. S.: BARC-705 (1973)
254. Verdizode, A. A., Alizode, T. D, Yusifova, G. N.: Ser. Khim Nauk No. *2*, 237 (1971)
255. Patel, B. K., Shah, N. H., Agarwal, Y. K.: Ann. Soc. Sci. Brux. Ser. *92*, 223 (1978)

256. Bykhovtsova, T. T.: Zavod. Lab. *43*, 1041 (1977)
257. Reddy, A. V. R. et al.: Radiochem. Radioanal. Lett. *47*, 45 (1981)
258. Davydov, A. V. et al.: INIS-mf-4047 (1977)
259. Davydov, A. V., Myasoedov, B. F., Tranikov, S. S.: Dokl. Akad. Nauk. SSSR *225*, 1075 (1975)
260. Fedoseev, E. V. et al.: Radiochem. Radioanal. Letters *51*, 271 (1982)
261. Dhingra, M. M., Subramanian, M. S.: Chem. Phys. Lett. *30*, 83 (1975)
262. Sekine, T.: Acta Chem. Scand. *19*, 1469 (1965)
263. Choppin, G. R.: Pure Appl. Chem. *27*, 23 (1971)
264. Ruben, J., Fiat, D.: J. Chem. Phys. *51*, 4918 (1969)
265. Allard, B.: J. Inorg. Nucl. Chem. *38*, 2109 (1976)
266. Pretorius, J. A., Boeyens, J. C. A.: J. Cryst. Mol. Struct. *6*, 169 (1976)
267. Pretorius, J. A., Boeyens, J. C. A.: J. Inorg. Nucl. Chem. *40*, 407 (1978)
268. Pretorius, J. A., Boeyens, J. C. A.: ibid. *40*, 1519 (1978)
269. Leipoldt, J. G., Bok, L. D. C., Lanbscher, A. E., Basson, S. S.: ibid. *37*, 2477 (1975)
270. Drew, M. G. B.: Coord. Chem. Rev. *24*, 179 (1977)

Surfactant Adsorption on Minerals Related to Flotation

Bohuslav Dobiáš

Department of the Physical Chemistry of Interfaces and Flotation, Institute of Physical and Macromolecular Chemistry, University of Regensburg, Regensburg, Federal Republic of Germany

Theoretical aspects of mineral flotation in particular the adsorption of flotation reagents at the mineral-water interface, were reviewed in 1959 for the last time[1)]. Several papers have been published since on this topic. The aim of this chapter is to survey recent publications that have contributed to the development of the flotation theory.

Structure and Bonding 56

1 Introduction

1.1 Flotation System and Basic Terms

Mineral flotation is a method for selective separation of mineral components out of polymineral dispersions of ground ores in water (ca. 5–35 vol.% of the solid) by using dispersed gas (usually air) bubbles. The method consists in the different adhesion of hydrophobized and hydrophilic mineral particles to an air bubble. Hydrophobized mineral particles adhere to the air bubble and are carried out as a specifically lighter aggregate to the surface of the mineral dispersion where they form a foam (froth) layer. This foam, called concentrate, is mechanically removed (Fig. 1 A). A mineral is hydrophobized by adsorption of a suitable surface-active compound (surfactant, collector) on the surface of the mineral component to be flotated. All other nonhydrophobized particles remain dispersed in the mixture (Fig. 1 B).

With this method it is possible to obtain highly pure mineral components from complex and especially from poor ores (even <1% of a useful component). The adsorption selectivity of the surfactant can be controlled by adding a depressant or an activator, by changing pH and/or the ionic strength of the solution. The foam layer must be sufficiently stable until the mineral particles are mechanically removed. If the surfactant is not stable enough, a frother must be added; this is mainly a nonionic surfactant. The physical chemical principles of flotation are studied in more detail by Schubert[2], Schulze[3] and Leja[4].

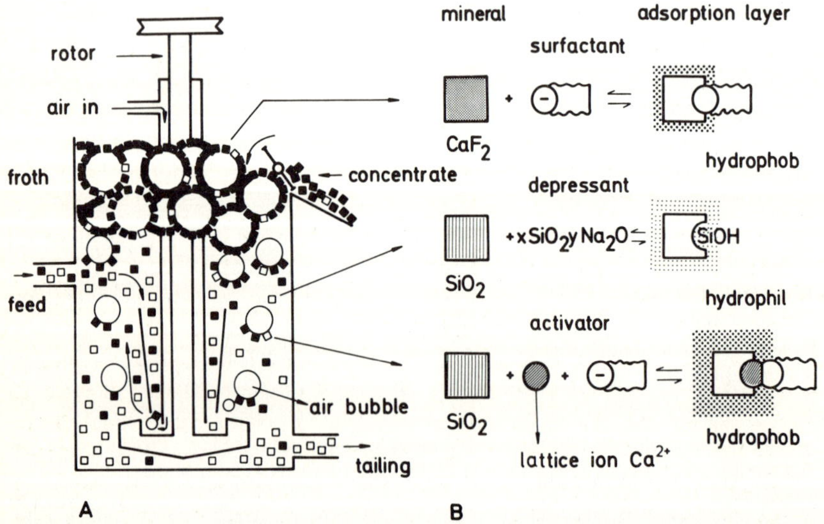

Fig. 1. A Principle of flotation in a mechanical-type cell. The rotor and stator (which is here omitted for simplicity) keep the mineral particles and air bubbles in dispersion for adhesion. **B** Formation of hydrophobic and hydrophilic adsorption layer on solid in quartz-fluorite system

2 Mineral-Water Interface

2.1 Character of Lattice Forces and Crystalline Surface

The surface properties of most minerals are generally influenced by the structure of the mineral-water interface and by surface species. The knowledge of these characteristics is required to understand the surface chemistry of solids. From the thermodynamic point of view, adsorption of surfactants on a mineral surface is very complicated. This is caused by the complex structure and ionic composition of the mineral-water interface formed during grinding the mineral in water.

Mineral grinding leads to distorsion of chemical and ionic bonds between atoms and ions. In the fracture areas binding and coordination states get asymmetric, and new electron and electric valences occur. Spontaneous reactions in the crystalline structure and with contact phases are the consequence of the distorsion. Surface distorsion of the crystalline structure may be diminished or completely abolished. At the same time, the free surface energy decreases due to polarization of surface ions. These ions are redistributed in the inner or outer layer of the crystalline surface and/or due to chemisorption of molecules and ions[1)]. All these changes occur side by side, but one of them can suppress the effect of the others in a decisive manner.

Beside the chemical composition, the crystalline structure of the mineral has an important effect on the adsorption ability of its surface. This is due to the fact that *lattice bindings* are usually not equivalent and space disproportions occur, so that fission surface areas have specific properties. Typical examples are layer lattices of graphite or talc where the main valences proceed in the layer plains whereas these are interconnected with feeble valences. Fission areas of such minerals are hydrophobic. The effect of the structure on adsorption properties of a mineral surface increases with increasing adsorption density and with decreasing force of the adsorption binding of the solid phase[5)]. A crystalline lattice contains structural defects (which include physical and chemical surface imperfections and deficiencies in the volume phase) which can influence the chemical reactivity of a crystal surface.

Lattice defects can function both as donors and as acceptors and create free electrons or electron holes. Crystalline surfaces containing unsaturated electron valences act as electron traps and capture free electrons. This leads to changes in binding conditions and in the charge state of e.g. metal ions: their ability to polarize O^- in a metal oxide decreases. Surface oxidation during the grinding process often causes deep alterations of the surface structure of solids (sulphides, graphite, coal). This usually leads to increases in affinity toward water and in reactivity with the surfactant.

Beside defects from mineral genesis, *grinding of a mineral* can produce roentgen amorphous states or a new crystalline phase. This leads to the formation of surfaces which differ morphologically and energetically from equilibrium surfaces. Relations were also observed between the degree of crystallinity and particle size on one side and surface reactivity with water or a surfactant on the other side. For example, the adsorption of xanthates on a very pure surface of pyrite monocrystals occurs much slower than on fine crystalline samples[5)].

The preparation of a mineral surface and its pretreatment have an effect on the adsorption of a surfactant because they can lead to a change in the stoichiometry of the surface ions. A typical final step in the preparation of mineral particles is *drying the mineral* in order to influence the character of its surface properties. This effect was studied by Dobiáš[6] and Kittabe[7]. Kulkarni and Somasundaran[8] examined the effect of leaching minerals with acids and bases on the electrokinetic properties of SiO_2 in aqueous solutions. Aging of the mineral surface as a result of phase equilibrium has also an effect on the surface properties of crystals[9].

2.2 Hydration, Electric Double Layer and Zeta Potential

Beside chemisorbed molecules of water on a mineral surface there is a special *water layer* of 50–200 nm thickness with an altered structure. Water molecules forming this layer are more structured in the direction toward the crystal surface. There are three layers of this "organized" (structured) water (Fig. 2). This so called interface (or vicinal) water is, according to the Drost-Hansen model[10, 11], independent of the formation of H^+-bridges or of the orientation of water molecules in the field of a polar crystal surface. On hydrophobic surfaces of solids (Fig. 2A) H-bridge bindings occur towards water in the bulk phase. The thickness of the layer of structured water corresponds to 10–100 water molecules and contains a very thin middle zone. On polar surfaces (Fig. 2B) the mode and extent of orientation of water molecules depends on the crystal structure and chemical composition of the solid, and includes the ion-dipol and ion-ion interactions as well as H^+-bridge bindings. With a rising dissimilarity between the polarity of surface and bulk water the thickness of the middle transient phase increases.

According to binding criteria, minerals can be divided into three groups:

a) minerals with a strong polar binding (oxides, silicates, salts),
b) minerals with an apolar binding (sulphur, coal, graphite) and
c) minerals with a weak polar binding (sulphides).

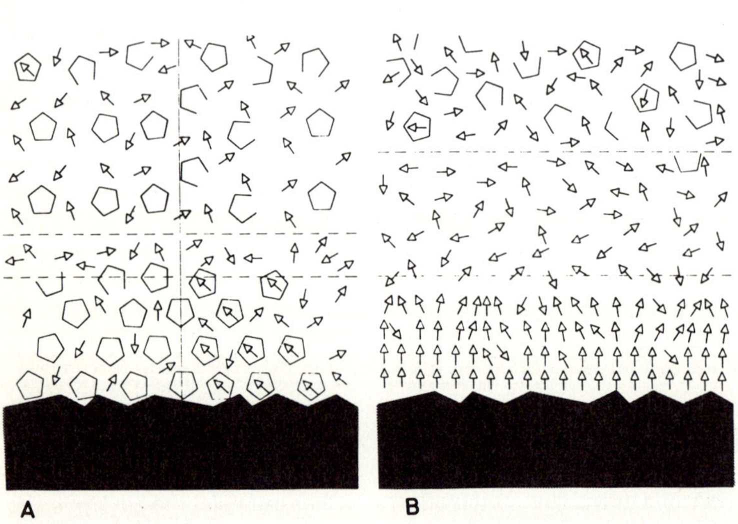

Fig. 2A, B. Three layers model of water at the interface with mineral according to Dorst-Hansen[164] ⬠ = clathrate-like ordering; ↑ = water-dipole. **A** Vicinal water near non-polar surface. Extensive clathrate-like ordering near interface but minimal disordered region. **B** Vicinal water near polar surface, disordered region

Wetting of a mineral with water generates a surface charge and a region of electrical inhomogeneity at the solid-water interface, as a result of the interaction of water molecules with the crystal surface. A surplus on positive or negative charge, which is typical for the crystalline structure of a mineral surface, is compensated by a region of counterions (H^+, OH^-, lattice ions etc.) of the opposite sign from the bulk phase (H_2O). Thus an *electric double-layer* (EDL) results at the mineral-water interface which in many cases has an important role in the adsorption of a surfactant. The EDL consists of a fixed and a mobile component. This present model of EDL results from Gouy and Champman's conception[12)] of the diffuse layer, that is characterized by the Poisson-Boltzman equation, and of a layer of specifically adsorbed ions according to Stern[14)] and Grahame[15)] (GChSG-model). This model assumes a firmly bound layer of adsorbed dehydrated (Grahame plane, or inner Helmholtz plane) and hydrated (Stern plane, or outer Helmholtz plane) ions on the electrically charged mineral surface, and a layer of free mobile hydrated ions which are on the boundary toward the bulk phase in a diffuse state (Gouy's layer). A schematic GChSG-model of an EDL is shown in Fig. 3, together with appropriate potentials[16)].

For an interpretation of the adsorption process it is important to know the so called *zeta-potential* (ζ) that can be calculated from electrokinetic measurements. It may be defined as the potential difference at the shear plane (near to the outer Helmholtz plane) between the diffuse layer outside the slipping plane and the bulk phase, when the solid and liquid phases are moved tangentially to each other. The location of the slipping plane is not exactly known, but it can be assumed that the shear plane is only very little further

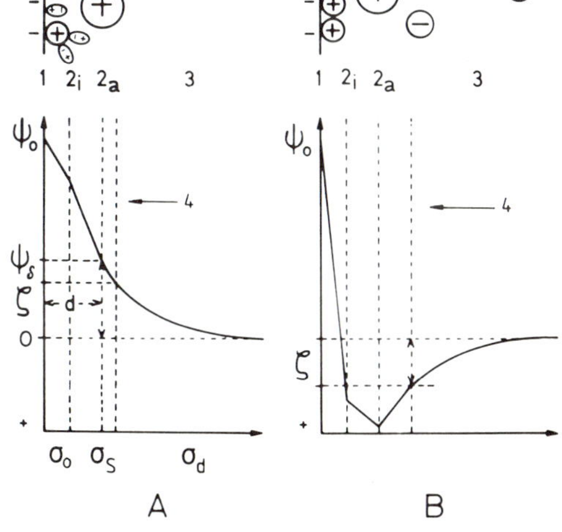

Fig. 3. The structure of the EDL at the mineral-water-electrolyte interface. 1-Layer of charging ions; 2_i-inner and 2_a-outer Helmholtz layer (Grahame and Stern plane, resp.); 3-diffuse layer and 4-slipping or shear plane [after Ref. 16]. ψ_0-phase potential and ψ_δ-Stern's potential. *a* – H_2O dipols, *b* – hydrated counter-ions, *c* – negatively charged ions, *d* – thickness of the G-S layer; σ – charge density

outwards form the surface than the outer Helmholtz plane; alternatively, it is located with in the Stern layer[1].

Hunter and Alexander[17)] located the slipping plane 1 nm from the kaolinite surface and Van Olphen gives a value of 0,5 nm for the thickness of the Stern layer on clay minerals[12)]. Furlong et al.[18)] suppose the distance of a few tenths of nm between the shear plane and the outer Helmholtz plane, so that $\zeta \equiv \psi_\delta$. The identity of ζ and ψ_δ is valid especially at low potentials (< 100 mV) and at low electrolyte concentrations ($< 10^{-2}$ mol · l^{-1}). At higher potentials and electrolyte concentrations the shear plane probably shifts further from the surface and ζ is expected to be less than ψ_δ. However, for the right interpretation of a ζ-potential it is necessary to consider all reactions in the EDL since different values of ψ_0 can correspond to one ζ-potential value. Changes in the ζ-potential will reflect changes in the surface potential only when the ionic strength remains constant and there is no specific adsorption other than by potential-determining ions, i.e. when only these and indifferent ions are present[19)]. For the study of electrokinetic phenomena, see e.g. the monograph of Show[13)].

Beside the classical GChSG model of the EDL there are some further concepts of its structure. Tadros and Lyklema[20, 21)] proposed a model based on a conception of a porous surface of the crystal. For example, an oxide surface may be porous to both potential-determining and counter ions, so that the high surface charge and a part of the counter-ion charge actually reside together inside the surface. Another porous model from Bérubé and de Bruyn[22)] postulates that the potential determining ions H^+ and OH^- do not adsorb on or in the mineral surface but remain in the structured water layer (separated from the solid surface).

The *porous double layer* (PDL), or gel layer, model was worked out quantitatively by Lyklema[23)] who assumed that the porosity and the surface charge decrease exponentially with increasing distance from the surface and that the adsorption of the counter-ions in the solid is controlled by a Langmuir-type equation. The theory and experiments are applied to the glass-solution interface. A different mathematical approach to the region of surface charge was used by Perram et al.[24, 25)] and by Levine et al.[26)] These authors assume a homogenous region of 2–5 nm thickness containing metal hydroxy groups $[Me(OH)_n]$ into which other ions are adsorbed and that the ion distribution in this gel layer corresponds to the Poisson-Boltzman equation. The fundamental difference between the two models, i.e. GChSG and PDL, is that in the former one the surface charge is assumed to be confined to a flat surface plane and the counter charge is contained in the solution side of the interface, whereas according to the latter model, the surface charge together with a part of the counter charge can penetrate into the solid[18)]. In the PDL models it is further necessary to distinguish between a physical porosity resulting from the presence of different openings, gel layer porosity due to the formation

1 To answer the question whether $\zeta = \psi_\delta$, Diemen and Stein[165)] compared the adsorption and electrokinetic potentials of mineral-water interfaces of some Ca silicates in aqueous solutions. The change of the potential in the chemisorption plane (locus of places with chemisorbed ions) with Ca^{2+} activity in the surrounding liquid is calculated from adsorption data. The potential, averaged over the Ca^{2+} sites, increases less with increasing Ca^{2+} activity then the ζ potential and the potential averaged over the whole chemisorption plane. In some cases, the distances between the electrokinetic slipping and chemisorption planes were calculated to be 1–2.5 nm. The distance would be too small if one considers that any surface disorder would increase the value, but it could also be overestimated if a lower value would be used for the dielectric constant

of a layer of hydrated species on a mineral surface, and finally chemical porosity resulting from the occupancy of metal ion sites by other ions of the same or smaller ion radius[25–27, 30)].

2.3 *Point of Zero Charge, Isoelectric Point and Surface Potential*

The formation of a charge on a mineral surface wetted with pure water is controlled by two processes:

a) adsorption of water molecules (dipols), H^+ or OH^- ions on the solid surface followed by a dissociation of the hydrolytic product formed;
b) preferential release of a certain lattice ion species from the solid phase as a result of different hydration energies and readsorption of hydrolytic lattice products.

The mechanism a) is characteristic for simple and complex metal oxides and a number of silicates whereas the mechanism b) is typical for sparingly soluble salts such as CaF_2, $BaSO_4$, $CaCO_3$. These mechanisms represent two extreme cases of a surface charge formation since in most minerals both mechanisms proceed simultaneously in proportions depending on the chemical composition and crystalline structure[31)].

Ions causing electrical charge of a mineral surface are called *potential-determining ions* (PDI). The surface charge density of a mineral σ_s according to the mechanism a) can be calculated from the change of pH value in an aqueous suspension of the solid by using Eq. (1):

$$\sigma_s = F(\Gamma_{H+} - \Gamma_{OH-}) \tag{1}$$

where Γ_{H+} and Γ_{OH-} are adsorption densities of H^+ and OH^- in mol · cm^{-2}, resp., as determined from acid-base titrations, and F is the Faraday constant. The rate of PDI adsorption is biphasic. The first adsorption step is fast and completed in a few minutes, the second step is slow and the equilibrium is reached in many days[32)].

Two important parameters describing the EDL of a mineral are the *point of zero charge* (PZC) and the *isoelectric point* (IP). Healy et al.[18)] define the PZC as the concentration of PDI with the surface charge of a mineral $\sigma_s = 0$ or $\psi_\delta = 0$. In metal oxides, PZC is determined by the concentration of PDI H^+ or OH^-, in sparingly soluble salts by the concentration of PDI of the lattice. When both mechanisms of surface charge formation operate simultaneously, both ion species and their reaction products determine the PZC[16, 31)]. The IP is defined[18)] as the concentration of PDI at which the electrokinetic potential $\zeta = 0$.

Samasundaran and Goddard[33)] cite the current *IUPAC definition* of these terms. Thus, PZC refers to a particle or surface carrying no fixed charge and IP to a particle showing no electrophoreses or a surface showing no electroosmosis. These are not interchangeable, since the IP characterizes the charge in the diffuse layer ($\sigma_d = 0$) whereas PZC should refer only to the surface charge. In the absence of a specific adsorption, the PZC and IP should coincide and are called intrinsic. Table 1 presents PZC an IP values of various minerals investigated. In the presence of unequal specific adsorption the PZC and IP move in opposite directions[34)]. Electrolytes in which the IP and PZC are identical and independent of the electrolyte concentration, are called indifferent and the values of

Table 1. Typical values of experimentally determined points of zero charge (PZC) and isoelectric points (IP) of some minerals

Mineral	IP/PZC pH	IP/PZC pMe^{n+}/pA^{n-}	Ref.
Quartz, SiO_2, different varieties	1.3–3.7		99
Quartz, Brasil	2.6		166
α-corundum, α Al_2O_3 freshly ground in air	8.2		6
Acid-leached (HCl 32%)	7.1		6
Heated at 700 °C after grinding	6.7		6
Grinded and stored in H_2O	9.1		44
Acid-leached and dried at 130 °C	6.7		167
boehmite, AlOOH	7.8–8.8		44
Hematite, Fe_2O_3, different varieties	5.4–8.7		99
Goethite, FeOOH	6.8		44
Rutile, TiO_2, different varieties	5.2–6.7		33, 44, 172
Zirconia, ZrO_2	4.0		44
Cassiterite, SnO_2	4.5		44
Cassiterite with 1.03% Fe_2O_3	3.95		60
with 6.86% Fe_2O_3	4.68		60
with 7.94% Fe_2O_3	4.80		60
Kaolinite, not washed	<3.0		169
8 × repeatedly washed in 1 M NaCl at pH 9	~3.7		169
Barite, $BaSO_4$, different varieties	3.3–5.3		166, 33, 170
		pBa^{2+} 3.9–7.0	33
		pBa^{2+} 6.7	44
Fluorite, CaF_2	8.25		166
Optically pure	10.75		32, 54
Another finding place	6.2	pCa^{2+} 2.6–7.7	33
		pCa^{2+} 3.0	44
Calcite, $CaCO_3$	9.5		33
Aged 30 min	5.5–6		33
Aged several days	8.2–8.5	pCa^2 3.5	33
		pCO_3^{2-} 3.0	33
Aragonite, $CaCO_3$	<5.0		33
Magnesite, $MgCO_3$, aged 30 min in KCl	6.0–6.5		33
Scheelite, $CaWO_4$		pCa^{2+} 3.1	171
		pWO_4^{2-} 6.2	171
		pCa^{2+} 4.8	44
Fluorapatite, $Ca_{10}(PO_4)_6F_2$	6.9 ± 0.2	pCa^{2+} 4.38	33
		pF^- 4.63	
		$pHPO_4^{2-}$ 5.2	
Leached with HNO_3	5.1		33
Leached with HCl	3.4		33
Fluorapatite, $Ca_{10}(PO_4)_6(FOH)_2$ equilibrated	~6.0		33
Nonequilibrated	3.75		
Hydroxyapatite, $Ca_{10}(PO_4)_6(OH)_2$	8.5 ± 0.2	pCa^{2+} 4.38	33
		$pHPO_4^{2-}$ 4.19	
Another finding place	7.0		44
Monazite (Ce, La, Th) PO_4	3.4		33
Another finding place	5.3–6.3		168

For IP and PZC of sulfides, see extensive data of Healy and Moignard[173)]

each term (IP, PZC) depend only on the surface properties of minerals and the composition of the aqueous phase. The indifferent electrolytes are adsorbed only by electrostatic attraction.

The value of the surface potential ψ_0 at any activity of PDI and a constant potential difference due to dipoles etc. is given by the Eq. (2)[35]:

$$\psi_0 = \frac{RT}{v_+ F} \cdot \ln \frac{(a_{Me^+})}{(a_{Me^+})_{PZC}} \quad \text{or} \quad \frac{RT}{v_- F} \cdot \ln \frac{(a_{A^-})}{(a_{A^-})_{PZC}} \tag{2}$$

where a_{Me^+} and a_{A^-} are activities of the positive and negative PDI in the solution, $(a_{Me^+})_{PZC}$ and $(a_{A^-})_{PZC}$ are their activities at PZC and v_+, v_- are the appropriate valencies. The charge in the diffuse double layer σ_a given by the Gouy-Chapman relation as modified by Stern for a symmetrical electrolyte is[35]:

$$\sigma_d = -\sigma_s [(2\,\varepsilon\, RT/\pi) \cdot c]^{1/2} \cdot \sinh (vF\psi_\delta/RT) \tag{3}$$

where ε is the dielectric constant of the liquid and c is the concentration of added electrolyte.

An experimental surface charge informs about the stoichiometry of the ionisation reactions and on the value of the ionisation constants. On the oxide surface holds:

$$AH_2^+ \rightleftharpoons AH + H^+ \tag{4}$$

$$AH \rightleftharpoons A^- + H^+ \tag{5}$$

$$AH_2^+ + Cl^- \rightleftharpoons AH_2Cl \quad \text{and} \tag{6}$$

$$A^- + Na^+ \rightleftharpoons ANa \tag{7}$$

where AH_2^+, AH, A^- represent positive, neutral, and negative surface sites, resp., and AH_2Cl and ANa the interfacial ion pairs; using NaCl as a 1:1 supporting electrolyte, Yates et al.[69] proposed a site-binding model in which the adsorbed counter ion form interfacial ion pairs with discrete charged surface groups. This model is used to calculate theoretical surface charge densities of PDI and the potential at the outer Helmholtz plane.

Davis et al.[70] describes a method for determining intrinsic ionisation and complexation constants of oxide surface sites from potentiometric titration data. Analyzing the local equilibrium reaction and the equilibrium constant of the surface sites they use the approach of Chan et al.[71]:

$$SOH_2^+ \overset{K_{a_1}^{int}}{\rightleftharpoons} SOH + H^+_{surf} \rightarrow K_{a_1}^{int} = \frac{[SOH][H_s^+]}{[SOH_2]} \tag{8}$$

$$SOH \overset{K_{a_2}^{int}}{\rightleftharpoons} SO^- + H^+_{surf} \rightarrow K_{a_2}^{int} = \frac{[SO^-][H_s^+]}{[SOH]} \tag{9}$$

and for NaCl

$$SO^- + Na_s^+ \underset{}{\overset{K_{Na^+}^{int}}{\rightleftharpoons}} SO^- - Na^+ \tag{10}$$

$$SOH_2^+ + Cl_s^- \overset{K_{Cl^-}^{int}}{\rightleftharpoons} SOH^+ - Cl^- \tag{11}$$

The $[H^+]$ at a location i in the EDL is related to the bulk solution concentration by the Boltzmann distribution $[H^+]_i = [H^+] \cdot \exp(-e\psi_i/kT)$. For concentrations of Na^+ and Cl^- holds:

$$[Na^+]_i = [Na^+] \cdot \exp(-e\psi_i/kT)$$
$$[Cl^-]_i = [Cl^-] \cdot \exp(+e\psi_i/kT)$$

The authors consider the reaction of Na^+ and Cl^- with the oxide surface as a complex ionization reaction and prefer the term "surface complexes" in contrast to Yates et al.[69] speaking on "ion pairs". It is therefore more instructive to write the reaction as complex ionization reactions:

$$SOH + Na_s^+ \overset{{}^*K_{Na^+}^{int}}{\rightleftharpoons} SO^- - Na^+ + H_s^+ \tag{12}$$

$$SOH + H_s^+ + Cl^- \overset{{}^*K_{Cl^-}^{int}}{\rightleftharpoons} SOH_2^+ - Cl^- \tag{13}$$

so that ${}^*K_{Na^+}^{int} = K_{a_2}^{int} \cdot K_{Na^+}^{int}$ and ${}^*K_{Cl^-}^{int} = K_{a_1}^{int}/K_{Cl^-}^{int}$.

When using these experimentally obtained figures and the stoichiometry of surface reaction, the surface charge adsorption density Γ_i and diffuse layer potential at the oxide-water interface can be calculated. In electrolyte solutions of moderate or high concentrations, the surface charge is dominated by surface complex formation of ionizable sites and electrolyte ions. In very dilute simple salt solutions the surface ionization reactions are more important. H^+ and OH^- are the main PDI, but the electrolyte ions have only a minor effect on the calculated surface potential ψ_0. The electrolyte ions and the PDI have a joint role in determining the magnitude of the surface charge via their reactions with the surface sites. The important experimental parameters for these contributions are

$$\Delta pK_{complex} = p^*K_{cation}^{int} - p^*K_{anion}^{int} \quad \text{and} \quad \Delta pK_a = pK_{a_2}^{int} - pK_{a_1}^{int} .$$

From the potentiometric titration data ($SiO_2 - H^+/OH^-$) Sidorova et al.[72] calculated the value of the surface potential ψ_0 and the dissociation constant of the surface silicic acid K_0 for the reaction:

$$\equiv SiOH \overset{K_0}{\rightleftharpoons} H^+ + \equiv SiO^- \tag{14}$$

$$pK_0 = pH + \log \frac{1-\alpha}{\alpha} + \frac{0.434 \cdot F}{RT} \cdot \psi_0 \tag{15}$$

where α is the degree of dissociation of the surface groups. Introducing the relationship pH vs. $\log(1-\alpha)/\alpha = pK$, where pK is the apparent dissociation constant, it is possible to write:

$$pK_0 = pK + \frac{0.434\,F}{RT}\,\psi_0 \tag{16}$$

The degree of dissociation is determined by the equation $\alpha = [\equiv SiO^-]/[\equiv SiOH] + [OH^-]$ for the surface groups in terms of the fraction of adsorption $\alpha = \Gamma_i/\Gamma_\infty$ where Γ_∞ is the maximum possible adsorption on quartz $8 \cdot 10^{-10}$ mol $\cdot$ cm^{-2} (assuming 0.255 nm^2 per a OH group). The surface potential is calculated according to the equation:

$$-\psi_0 = \frac{2.3\,RT}{F}\left(pH + \log\frac{\Gamma_\infty - \Gamma_i}{\Gamma} - pK_0\right) = \frac{2.3\,RT}{F}\,(pK - pK_0) \tag{17}$$

The value of pK_0 is determined by extrapolating curves plotted on coordinates of pK vs. α to a zero value of α (uncharged surface). The value found for $pK_0 = 5.3$ proved to be considerably lower than pK of free silicic acid in the solution. When an ion-permeable "gel layer" is present, Gouy's equation to calculate ψ_0 from adsorption data is not valid. Detailed measurements and calculations on γ-Al_2O_3 showed that the calculated values of ψ_0 agree well with values of ψ_1 calculated according to Gouy's equation:

$$\Gamma = \frac{1}{F}\sqrt{\frac{\varepsilon\,RTC}{2\pi}} \cdot 2\sinh\left(\frac{F\psi_1}{2RT}\right) \tag{18}$$

2.4 *Charging Mechanism*

The oxide-water interface with TiO_2, SiO_2, Al_2O_3 or Fe_2O_3 minerals is most often studied to clarify how the surface charge is formed. As shown e.g. by Boehm et al.[36–38] during hydration of the (TiO_2) surface the coordination gaps of the crystal are filled due to a movement of protons toward surface O^{2-} ions. At the same time the electrostatic potential decreases. Since the Me-O binding in TiO_2 and in other metal oxides is not exclusively ionic, but has a covalent character, it is possible to speak of surface OH groups. Chemisorption of water molecules on the metal oxide surfaces induces a formation of surface OH groups with an amorphous character (pH 6.3–6.6); the dissociation of these groups leads to an acidic or alkaline surface (positive or negative surface charge)[38]. Measurements on a freshly ground and hydroxylated silica showed two types of OH-groups: isolated or vicinel and paired or geminal ones[39–41]:

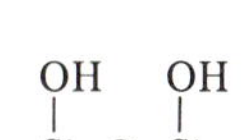

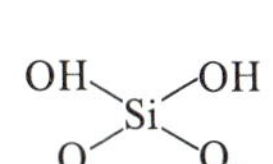

(19)

vicinal hydroxy groups geminal

Hockey et al. found that the surface density of isolated and paired OH-groups is 1.4 ± 0.1 and 3.2 ± 0.1 per nm^2, resp.[42]. The value depends on the underlying surface ions and can vary with the crystal plane at the surface. IR-spectrometric measurements with TiO_2 (rutile) vevealed on the metaloxide surface both types of OH-groups and water molecules that are bound to the surface by coordination forces. The theoretical surface density of OH groups and chemisorbed H_2O was also calculated from the lattice parameters of the surface planes. For example, the rutile plane 110 contains 60% of both OH groups types and the surface planes 101 and 100 contain 40% of water bound by coordination forces[43].

Dehydration of the surface of e.g. SiO_2 leads to the formation of siloxan groups Si-O-Si and is reversible up to the temperature of ca. 450 °C. The octahedral surface of β-cristobalite may contain 6.2–6.7 OH groups/nm^2 at 25 °C; after dehydration at 150 °C, the number decreases to 5 OH groups/nm^2, to 4.6 OH groups at 450 °C and to 1 group at 800 °C/nm^2, provided one surface Si atom contains only 1 OH group[43]. When silica is heated, surface silanol groups start to condense and to eliminate water at about 170 °C. Above 850 °C no chemisorption of water is possible and the surface becomes hydrophobic.

From the existing charging mechanisms we mention the most often used adsorption-dissociation mechanism of Fuerstenau and Healy[44] and of Stumm et al.[62]

$$MeOH_{(surface)} \rightleftharpoons MeO^-_{(surface)} + H^+_{aq} \tag{20}$$

$$MeOH_{(surface)} + H^+_{aq} \rightleftharpoons MeOH^+_{2\,(surface)} \tag{21}$$

An alternative charging mechanism was proposed by Parks and de Bruyn[45] who postulate the adsorption of charged hydroxo complexes of the form $Me\,(OH)^{z-m}_m$:

$$Me_2O_{3\,(solid)} + 3\,H_2O \rightleftharpoons 2\,Me\,(OH)_{3\,(aq)} \tag{22}$$

$$Me\,(OH)_{3\,(aq)} \rightleftharpoons Me\,(OH)^{3-m}_{m\,(aq)} + (3-m)\;OH^-_{(aq)} \tag{23}$$

$$Me\,(OH)^{3-m}_{m\,(aq)} \rightleftharpoons Me\,(OH)^{3-m}_{m\,(surface)} \tag{24}$$

Van Olphen[46] and Parks[47] assume that the surface charge forms due to lattice substitutions. In clay minerals, e.g., the substitutions of Al^{3+} for Si^{4+} and Mg^{2+} for Al^{3+} produce a net negative charge in the clay lattice that is independent of the conditions in solution. This charge dominates the sheet surfaces of the clay minerals. In contrast to the edges, a "rupture bond" structure with potential determining ions H^+ and OH^- is not present on the sheet. The combined result of the rupture bond at edge surfaces and of negatively charged sheets on kaolinite is a mineral particle with the PZC of about pH 3 to 5, and its edges may be positive[48] even at a value of pH 9[19].

The effect of different Al : Si and O : Si ratios in aluminosilicates on the formation of the surface charge was studied by Smolik et al.[49] and Deju and Bhappu[50, 51], resp. The found that *acid leaching* led to a preferential removing of Al^{3+} from the surface and reduced the PZC. The PZC values increased with increasing the O : Si ratio. In acid solutions, the ability of H^+ to substitute for Li^+, K^+, Na^+, Ca^{2+}, and Mg^{2+}, grows with increasing O : Si ratio and enhancing Fe content.

Sparingly soluble salt-type minerals contain lattice ions as PDI but also H^+ and OH^- and their complexes if the latter resulted from hydration equilibrium between the crystalline surface and water[31]. For example, when calcite reacts with water, HCO_3^-, CO_3^{2-}, Ca^{2+}, $CaHCO_3^+$, and $CaHO^+$ are formed as surface charge species[52]. At high pH values, the calcite surface will contain a surplus of HCO_3^- and CO_3^{2-} and will carry a negative charge whereas at low pH it will have surplus of Ca^{2+} and $CaHCO_3^+$ ions and a positive charge. These surface ions are formed on the mineral-water interface or in the solution and adsorbed on a solid according to their concentration in the solution. Somasundaran and Agar[53] consider OH^-, H^+, and HCO_3^- as PDI in addition to Ca^{2+} and CO_3^{2-}. Since the concentration of all the species can be calculated as a function of pH from the available thermodynamic data, the PZC of the mineral is obtained assuming a correspondence to the IP of the solution, by estimating the pH value at which the total concentrations of negative and positive PDI are equal.

PZC determined from electrokinetic measurements at pH 8.2 agrees well with PZC values obtained from measurements of the minimum solubility and flotability of calcite[53].

The mechanism of *surface charge formation* on fluorite was studied by e.g. Dobiáš[31, 54]. Adsorption measurements in pure water at pH 6.1 ± 0.1 showed that OH^- ions adsorb on the CaF_2 surface to an extent ($2.8–6.1 \cdot 10^{-15}$ mol · cm^{-2}) which is substantially smaller than the amount of PDI of H^+ on the surface e.g. of positively charged α-corundum ($1.1–1.8 \cdot 10^{-12}$ mol · cm^{-2}). The PZC and IP of CaF_2 were found to be pH 5.8 and 10.75, resp. In spite of OH^- adsorption, CaF_2 has a high positive charge in pure water with a maximum in the region of PZC. An increase in H^+ concentration enhances Γ_{H^+}, but does not influence the positive value of ζ potential. In agreement with findings of Shergold[55], the increase in H^+ concentration on a CaF_2 surface is caused by a substitution of H^+ for Ca^{2+} and, at pH 3.25, by the formation of surface molecular HF. In a strongly acidic solution the PDI of fluorite are H^+ and Ca^{2+}, whereas Ca^{2+}, $Ca(OH)^+$ and F^- play this role in alkaline solutions. The PDI in pure water are mainly Ca^{2+} and F^-, when H^+ and OH^- participate only to a negligible extent.

The charge formation on the surface of tri-ionic crystals (apatite) is still more complex. Samani et al.[56] postulate Ca^{2+} and HPO_4^{2-} as PDI of fluoroapatite; pH controls the concentration of these ions in the solution and the hydrolysis rate of appropriate surface species. In this way H^+ and OH^- indirectly become PDI. Dobiáš et al.[57] came to the same conclusion, but they also include F^- as PDI. Somasundaran[58] takes H^+, OH^-, and phosphate ions for major PDI; Ca^{2+} and F^- have a major influence on the ζ potential of fluoroapatite.

The valence of both the anion of an acid and the cation of a base have a substantial influence on the character of the surface charge and on the value of PZC and IP[59]. PZC and IP of a mineral are further affected by the mineral's genesis, by a preparative pretreatment of its surface with chemical agents, the presence of impurities (other minerals), isomorphism, temperature pretreatment, and age of the surface[58, 60, 61].

2.5 Predicting the Point of Zero Charge

There are two methods to predict the PZC of oxides and hydroxides. The first one is based on the theory of minimum solubility of the mineral according to which the PZC should be identical to the IP and is determined by constructing the solubility diagram[63, 64]. The other method uses Park's equation[64, 65] IP = $A_{eff} - 11.5 \times [Z/R + 2.9 \times 10^{-3} x (CFSE)]$, where Z is the formal charge of the cation and R is the sum of cationic radius (r_+) and the oxygen diameter (0.28 nm). CFSE is the crystal field stabilization energy and A arises from changes in coordination number of hydration. From Park's equation and Pauling's electrostatic valence principle, Yoon, Salman and Donnay[66] derived a relation which describes the surface charging mechanism of oxide and hydroxide minerals in aqueous media and predicts the PZC from the crystallographic data.

By using the total free energy change (ΔG^0) and the equilibrium constant K of the protonization reaction $Me\text{-}O^{(2-v)-} + 2H^+ \rightleftarrows Me^{v+} + OH_2$, the authors developed the following expression for predicting PZC:

$$PZC = A - B\left(\frac{v}{L}\right) - \frac{1}{2}\log\left(\frac{2-v}{v}\right) \tag{25}$$

where v is the bond valence reaching the O^{2-} and $L = \bar{L} + r$. $\bar{L}$ is the mean Me-O bond distance in the interior of a crystal and r may be 0.101 nm for the distance O–H in water, hydroxides, and hydrated solids. The values of $\bar{L}$ are published by Donnay et al.[67] and Wyckoff[68].

Although the terms A and B from Eq. 1 contain ΔG^0_{chem} and effective dielectric constants, the values of which are not known at present, it is possible to determine the terms by using the experimental PZC values of solids. For their calculations the PZC's of α-Al_2O_3 (pH 9.1) and MgO (pH 12.4) were used, so that it may be written:

$$PZC = 18.43 - 53.12\left(\frac{v}{L}\right) - \frac{1}{2}\log\left(\frac{2-v}{v}\right) \tag{26}$$

The two oxides require no correction for CFSE, and the Me-O bondings are predominantly ionic. A linear relationship between PZC + 1/2 log (2 – v/v) and (v/L) confirms the plausibility of the Eqs. 25 and 26 for predicting PZC's in a series of oxides. A certain deviation behaviour of hydrogen-bonded hydroxides and oxyhydroxides from the linear relationship is explained by the fact that structure-forming OH^- have a specific affinity toward the broken H-bonds on the surface layer and may be classified as "non-dissociative adsorption".

Oxides of transition metal compounds need special consideration for CFSE since d-electrons are involved in chemical bonding. Then Eq. 25 may be written:

$$PZC = 18.43 - 53.12\left[\left(\frac{v}{L}\right) + \frac{c}{53.12}\cdot C\right] - \frac{1}{2}\log\left(\frac{2-v}{v}\right) \tag{27}$$

in which C is CFSE and c/53.12 is a constant to be determined from the value of experimental PZC of $Co(OH)_2$ (pH 11.4). C of $Co^{2+} \times 6H_2O$ has the value of 138 kJ/mol. Eq. 27 may also serve for calculating the Gibbs free energy of H^+ adsorption at PZC:

$$\Delta G^0 = -2.3\,RT\left[PZC + \frac{1}{2}\log\left(\frac{2-v}{v}\right)\right] \tag{28}$$

Table 2. Calculated and experimental PZC's of oxides and hydroxides after[66)]

Sample	Z	CN Me	CN (O, OH)	L	ν	(ν/L)	CFSE kcal/mol	$(\nu/L)_{eff}$	PZC Calculated	PZC Experimental
MgO	2	6	6	2.106	1/3	0.1070	–	–	12.40	12.4 ± 0.3
La_2O_3	3	7	4	2.44	1/2	0.1449	–	–	10.49	10.4
ZrO_2	4	7	4	2.26	1/2	0.1529	–	–	12.07	10–11
BeO	2	4	4	1.65	1/2	0.1880	–	–	8.20	10.2
ZnO	2	4	4	1.978	1/2	0.1673	–	–	9.30	9.3
α-Al_2O_3	3	6	4	1.91	1/2	0.1712	–	–	9.10	9.1
$Al(OH)_3$*	3	6	2	1.89	1/2	0.1724	–	–	9.03	5.0–5.2
α-AlO(OH)	3	6	–	1.935	1/2	0.1698	–	–	9.17	7.7
α-Fe_2O_3	3	6	4	2.03	1/2	0.1645	–	–	9.45	9.04
α-FeO(OH)	3	6	–	2.04	1/2	0.1639	–	–	9.48	6.7
ThO_2	4	8	4	2.425	1/2	0.1456	–	–	10.46	9.0–9.3
PuO_2	4	8	4	2.336	1/2	0.1492	–	–	10.27	9.0
Y_2O_3	3	6	4	2.28	1/2	0.1520	–	–	10.12	9.0
SnO_2	4	6	3	2.053	2/3	0.2177	–	–	6.71	6.6
TiO_2	4	6	3	1.969	2/3	0.2245	–	–	6.35	6.7
SiO_2	4	4	2	1.61	1	0.3817	–	–	-1.85	1.8
$Fe(OH)_2$	2	6	3	2.11	1/3	0.1068	18	0.1169	11.87	12 ± 0.5
$Co(OH)_2$	2	6	3	2.10	1/3	0.1072	33	0.1257	11.40	11.4
$Ni(OH)_2$	2	6	3	2.04	1/3	0.1093	42	0.1329	11.02	11.1
CuO	2	4	4	1.95	1/2	0.1689	39	0.1908	8.06	9.4 ± 0.4
$Cu(OH)_2$	2	4	2	1.935	1/2	0.1698	39	0.1917	8.01	7.7

PZC of alumosilicates (kyanite, silimanite, andalusite) can also be predicted from crystallographic data when we consider them as complex oxides:

$$PZC = \frac{\sigma_i}{K} + 18.43 - 53.12 \sum_{i=1}^{n} f_i \left(\frac{v}{L}\right)_{eff}^{i} - \frac{1}{2} \log \sum_{i=1}^{n} f_i \log \left(\frac{2-v}{v}\right)_i \tag{29}$$

where σ_i is the intrinsic structural charge of the component i, f_i is the atomic fraction on the surface and n is the distinct component of oxides.

The calculated and experimentally found values of PZC's of several oxides and hydroxides are compared in Table 2[66)].

3 Basic Adsorption Hypotheses and Isotherms

3.1 Langmuir and Phase Transition Hypothesis

Various hypotheses were conceived for the study of elementary processes during adsorption on solids. For surfactant adsorption, the most important ones are described by Richter and Schneider[74)].

According to the Langmuir conception, there are certain preferential adsorption sites on a mineral surface called *active centers*, but they are not exactly specified. Once adsorbed, the particles remain firmly bound. The active conters are at such distances that the adsorbed particles cannot affect each other. This hypothesis agrees well with conceptions of crystal formation where steps, edges, corners, shifts, and atomic defects can act as active centers. The conception of a real crystalline structure according to the Monte-Carlo method is shown in Fig. 4[75]. It is known that the surface diffusion is faster along surface imperfections than at an intact lattice. The lattice geometry of a crystal surface has no influence on adsorption. The presence of *defects on the surface* and preferential adsorption sites was demonstrated e.g. by Schneider et al.[76, 77]. Although the importance of structural defects, especially of shifts, for adsorption was clearly shown it has not yet been possible to develop a quantitative relationship between structural defects and the adsorption.

Models of phase transition are based on the conception that, because of an arranged lattice structure, a crystalline surface is a carrier for a periodic potential field at the atom lever so that the adsorbate is exposed to various power effects when it is with in the effective sphere of the field. The difference from the Langmuir model also consists in a postulated interaction between the adsorbed molecules. As a result of the loss of energy in the potential field of the adsorbent, the adsorbed particles can move only in two-dimensions. This phenomenon is very important for the adsorption process and is called *Volmer diffusion*[78]. According to Volmer, the number of particles moving to adsorption sites by diffusion is greater than the number of those bound to a certain site or those leaving the crystalline surface.

If the mutual interaction of adsorbed particles is strong enough the *surface diffusion* can become organized and leads to a formation of islets of adsorbed molecules in the sense of the Volmer-Weber "drop" mechanism, or to a layer association, parallel to the crystalline surface, according to the Frank-van der Merve's mechanism. Alternatively, it can form a tridimensional organized region (Stranski-Krastanow's mechanism) similar to that of adsorption mixed crystals[74].

If the adsorption results in an organized deposit layer and a certain arrangement similar to the crystalline lattice we speak of epitaxy. A prerequisite of epitaxy is not only a two-dimensional lattice analogy of foreign lattice layers but also the mode of nucleus formation, fusing the nuclei, purity of crystal surface, and binding energy between lattice

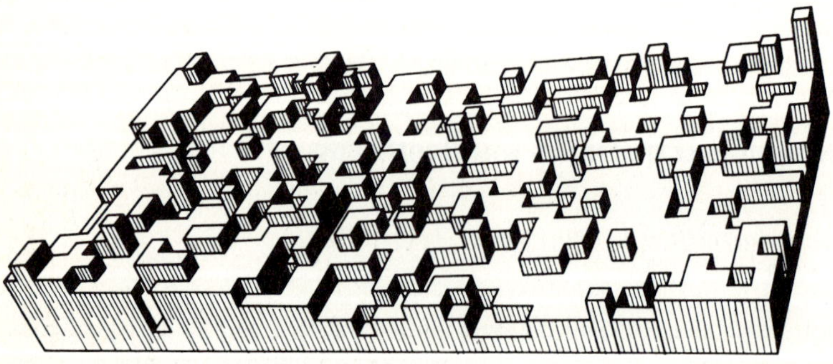

Fig. 4. Model of the real crystalline surface using the Monte-Carlo-method [after Ref. 75]

elements and the adsorbed particle. Epitaxy also occurs as a result of large misfits in distances between lattice layers growing together because this difference is counterbalanced by a special shift arrangements in the inter-layer[79)].

According to the kind of interaction forces and binding relationship between the solid and the adsorbed particle, we distinguish the physical adsorption and chemisorption. The real state of most systems is between these two extremes. For a detailed study of both adsorption mechanisms see e.g. Fripat et al.[80)].

The *physical adsorption* is characterized by weak intermolecular forces of the van der Waals type. The adsorbed particle must get close to the solid surface, since the van der Waals energy is proportional to the sixth power of reciprocal distance. The main feature of this interaction is its non-specificity, a considerable velocity and reversibility. An example of the physical adsorption is the adsorption of apolar molecules on an apolar surface resulting form disperse forces. Beside these forces the dipol-dipol interactions occur when molecules of the adsorbent or adsorbate can form permanent or induced dipoles (adsorption of gases or dipol liquids on apolar surfaces).

Chemisorption results form interaction of electrons or their transitions and is characterized by the formation of instabile surface compounds as precursors of chemical binding. It is almost limited to surfaces with a metal or ionic lattice. Its main features are its specificity, a relative slowness, need of a more or less high activation energy, and irreversibility.

Beside the described interactions, an adsorption of apolar and polar molecules (*ion-dipole interaction*) occurs on the surface of ionic crystals. The obtained energy is greater than the van der Waals energy, but smaller than the Coulomb interaction. Still stronger interaction can be expected during ion adsorption on ionic crystals. If ions adsorb directly on species of crystalline surface, the energies derived from the Coulomb's interaction are comparable to energies of the chemical binding.

3.2 Classical Adsorption Isotherms of Langmuir, Freundlich and Těmkin

Adsorption isotherms represent a relationship between the adsorbed amount at an interface and the equilibrium activity of an adsorbed particle (also the concentration of a dissolved substance or partial gas pressure) at a constant temperature. The analysis of adsorption isotherms can yield thermodynamic data for the given adsorption system. Theoretical adsorption isotherms derived from statistical and kinetic data, and using the described assumptions (see 3.1), are known only for the gas-solid interface or for dilute solutions of surfactants (Gibbs). Those for the system gas-solid are of a few basic types that can be thermodynamically predicted[81)]. From temperature relations it is possible to calculate adsorption and activation energies or rate constants for individual isotherms. Since there are no theoretically founded equations of adsorption isotherms for dissolved surfactants on solids, the adsorption of gases on solides can be used as a starting point for an interpretation.

The Langmuir adsorption isotherm is the most frequent relationship for such interpretations. From a series of various described forms we chose that according to Richter and Schneider[74)]. A prerequisite of the Langmuir-type course of the adsorption isotherm is that the adsorption leads to a monomolecular cover of the particular surface, that it is localized, that all sites to be occupied are energetically equal, and that the adsorbed particles do not interact with each other[4, 81].

For the adsorbed amount m one can write:

$$m = m_{max} \frac{c}{\beta + c} \quad \text{or, in a linearized form,} \quad \frac{1}{m} = \frac{\beta}{m_{max} \cdot c} + \frac{1}{m_{max}} \qquad (30)$$

where m is the molality of the substance adsorbed (m = number of moles/adsorbent mass), m_{max} is the maximum amount that can be adsorbed at a monomolecular coverage and c is the concentration of the adsorbed compound in the solution. β is the Langmuir constant and is equal to $\beta_0 \cdot \exp(+E/kT)$ were β_0 is the factor of proportionality and E ist the adsorption energy. For very low concentrations of the dissolved substance ($c \ll \beta$) the Langmuir equation gives a linear relationship according to Henry:

$$m = m_{max} \cdot \frac{c}{\beta} \qquad (31)$$

When $c \gg \beta$, the isotherm resembles the monomolecular coverage and $m = m_{max}$.

The linearized Eq. 30 is used to verify the Langmuir adsorption mechanism. The constant β can be estimated from the slope of regression lines. For a two-point adsorption, when a molecule adsorbs after a former dissociation at the interface, Eq. 30 gets the form

$$m = \frac{m_{max}\sqrt{\beta \cdot c}}{1 + \sqrt{\beta \cdot c}} \quad \text{or, in a linearized form,} \quad \frac{1}{m} = \frac{1}{m_{max} \cdot \sqrt{\beta}} \cdot \frac{1}{\sqrt{c}} + \frac{1}{m_{max}} \qquad (32)$$

For an energetically heterogeneous surface, where centers of different adsorption intensity are scattered, a superposition of the Langmuir isotherm results in

$$m = \sum_i \frac{m_{max\,i} \cdot c}{\beta_i \cdot c} \qquad (33)$$

In this case it is difficult to separate isotherms that belong to certain centers. Adsorption isotherms of an empirical character are described by Freundlich's equation[4, 81, 82]:

$$\frac{k}{k_0} = a \cdot \left(\frac{c}{c_0}\right)^n \quad \text{or as a linearized form} \quad \log\frac{k}{k_0} = \log a + n \cdot \log\left(\frac{c}{c_0}\right) \qquad (34)$$

where k is the amount adsorbed, k_0 and c_0 are units used for k and c, resp., so that a and n are temperature-dependent constants that characterize the given system. The validity of the isotherm is confirmed by a linear course of the relationship log k vs. log c.

For an inconstant adsorption energy an adsorption equation was developed by Těmkin for isotherms[83] where this energy decreases with increasing occupation of the surface by the adsorbed molecules:

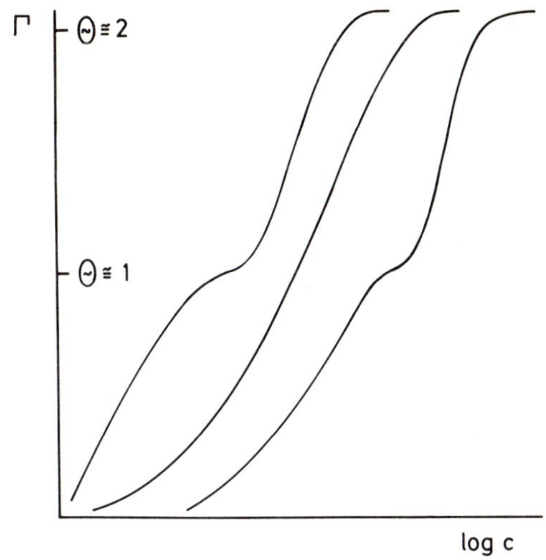

Fig. 5. Shapes of the adsorption isotherms of surfactants on minerals, which can be described using Langmuir or Těmkin equation for range $\Theta \leqq 1$

$$m = m_1 \ln(A \cdot c) \tag{35}$$

where m_1 and A are constants. The temperature-dependent constant A can be expresses as $A = A_0 \cdot \exp.(-E_0/kT)$ for a very low accupancy of the surface E_0. The Těmkin's adsorption can be verified by plotting m against ln c.
Dobiáš[31, 84–87] interpretated the adsorption isotherms of surfactants on the surface of minerals with a different character of PDI by using the described equations. These adsorption isotherms have in most cases a S-or double-S-form (Fig. 5) and can be described with the Langmuir or Těmkin equation. The adsorption of the surfactant conforms to the Langmuir equation in the concentration range close to the monomolecular coverage ($\Theta \cong 1$) on minerals with PDI H^+ and OH^-. For adsorptions on mineral surfaces with lattice ions as PDI the mentioned equation is less suitable because it is valid only for lower surfactant concentrations. Here the Těmkins equation fits well in the range of intermediate coverage values ($\Theta = 0.2$–0.8). Freundlich's adsorption equation can be used in both cases only at the start of the adsorption for very low surfactant concentrations.

3.3 Adsorption Isotherms with Terms for Lateral Interactions

Adsorption isotherms for mobile adsorbates were proposed by de Boer[4, 88] who though assuming lateral interactions between adsorbed molecules stays at the model of energetically uniform adsorbent surface. Starting from the analogy of the van der Waals equation for a three-dimensional system, de Boer developed an equation for a non-ideal behaviour of adsorbate in a two-dimensional adsorbed film:

$$\left(\phi + \frac{\alpha}{A^2}\right)(A - \beta) = kT \tag{36}$$

where ϕ is the film pressure $(\gamma - \gamma_0)$, A is the area of the film and α represents the lateral interactions between adsorbates. β characterizes the co-area of the adsorbate molecule and is equal to $1/2\pi \cdot d^2$ (d is the diameter of the adsorbate molecule).
By using the Gibbs adsorption theorem and the relation between the three-dimensional pressure p and the film pressure ϕ de Boer and Hill derived from Eq. 36 and equation for a two-dimensional adsorption with two separate types of interactions[4]:

$$p = K \frac{\Theta}{1-\Theta} \exp\left(\frac{\Theta}{1-\Theta} - \frac{2\alpha\Theta}{kT\beta}\right) \tag{37}$$

where Θ is the surface coverage by adsorbed molecules. The parameter K represents interactions between the adsorbate and the adsorbent and the term $2\alpha/kT\beta$ represents mutual interactions between the adsorbate molecules. This so called Hill de Boer equation permits the interpretation of the two-dimensional phase change in the adsorbed film during adsorption (condensation, order-disorder etc.) Cases et al.[89–92] developed adsorption isotherms for homogeneous and inhomogeneous mineral surfaces on the basis of surfactant adsorption on solids. For homogeneous surfaces they start from the common equation for localized adsorption without dissociation of adsorbed molecules[89]:

$$kT \cdot \ln x = -kT \cdot \ln W_a - \phi_a + kT \cdot \ln \frac{\Theta}{1-\Theta} + A \tag{38}$$

where x is the molar fraction of the solution in an equilibrium with the adsorbed layer at the coverage Θ (Θ = total amount/maximum amount of adsorbed surfactant molecules per a unit area), $k \cdot \ln W_a$ is the sum of all entropic terms for a molecule in the adsorbed phase, except the configuration entropy equal to $k \cdot \ln \Theta/1 - \Theta$, ϕ_a is a differential energy of desorption per molecule, and A is a constant. From this equation Prédali and Cases[89] derived the Frumkin-Fowler isotherm[93, 94] for a case that ϕ_a is dependent on the surface coverage Θ. In a reverse case one can arrive to the Langmuir adsorption isotherm. The mentioned authors assume that the term ϕ_a consists of two members of which the first one corresponds to a normal bonding between one adsorbed molecule and the surface, and the second one includes the lateral interaction between molecules in the adsorbed layer, assumed to be proportional to Θ $(\phi_a = \phi_a^0 + \Theta\omega)$. ω is therefore the sum of the lateral bindings between two neighbouring molecules when $\Theta = 1$. The molar fraction x can be further substituted with the molar fraction x_0 for the equilibrium with the micelles of a surfactant $(\Delta\mu = kT \cdot \ln x/x_0)$; moreover it holds:

$$kT \cdot \ln x_0 = -kT \cdot \ln W_0 - \phi_0 + A \tag{39}$$

where ϕ_0 is the energy of dissolution of the micelle, equal to the work per molecule needed to dissolve the micelle when molecules from the half-crystal position are broken away; $kT \cdot \ln W_0$ is the sum of all entropic terms for a molecule in the half-crystal position[95].

Provided the number of coordinations in the adsorbed layer and the lateral interactions are the same as in a plane of the lattice of the micelle, ϕ_0 can be expressed by:

$$\phi_0 = \phi_0^0 + \frac{\omega}{2} \tag{40}$$

where ϕ_0^0 is half of the normal energy for a molecule of a lattice plane from the half-crystal position with other molecules outside this plane. If $W_a \cong W_0$ are the same in the adsorbed layer and in the micelle, the following expression can be obtained from the above mentioned relations for the thermodynamic equilibrium between the adsorbed surfactant layer and its solution:

$$kT \cdot \ln \frac{\Theta}{1-\Theta} = (\phi_a^0 - \phi_0^0) - \frac{\omega}{2}(1 - 2\Theta) + \Delta\mu \qquad (41)$$

where $\Delta\mu$ characterizes the undersaturation of the surfactant in the solution. The course of the *Frumkin-Fowler isotherm* according to Eq. 41 is shown in Fig. 6 A. This isotherm has a point of inflection for $\Theta = 0.5$. If $\omega > 4\,kT$ (10 kJ/mol at 25 °C), the slope of the curve 1 is negative. The part of the curve between M and N is related to a change of state due to the condensation of the layer. When $\omega < 4\,kT$ (for short alkyl chains) the equilibrium concentration increases without a phase change (curve 2). Moreover, the adsorbed layer condenses and the micellar layers are no more compact. The equilibrium between two states of the layer occurs for $\Delta\mu = \Delta\mu^*$. Increasing chain length displaces the isotherms along the abscissa due to the variation of desorption energy for a surfactant molecule from the half-crystal position of the two-dimensional condensed phase. This shift is independent of the normal energy of desorption ϕ_0^0. The transition of the adsorbed layer from $\Theta \ll 1$ to a condensed state where $\Theta \cong 1$ can occur with only one kind of active sites.

When the crystal surface contains sites of different adsorption energies the adsorbate will condense area by area. According to Prédali and Cases[89] the adsorption isotherm will be steplike in relation to condensation of molecules on areas i, and each step i is compared with sites having the same energy $\phi_{a,i}^0$. Developing the isotherm with condensation on inhomogenous surfaces they start with the definition of the fraction of sites α and β on which condensation occurs between undersaturation $\Delta\mu$ and $d\Delta\mu$. When all the

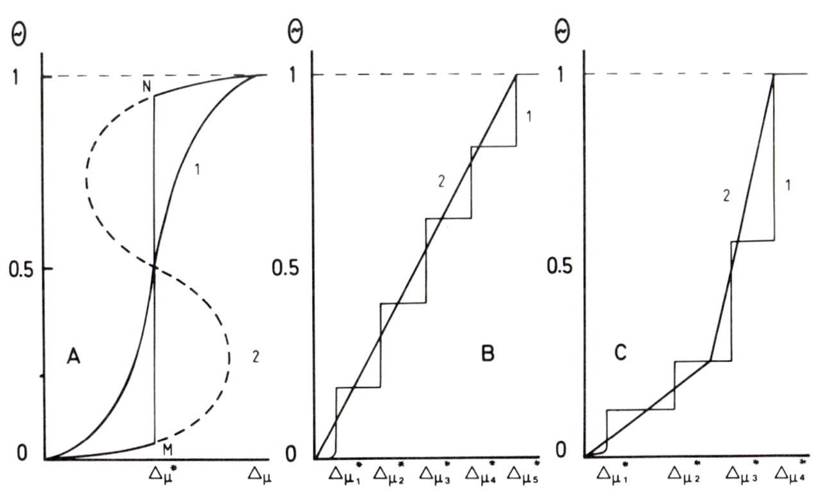

Fig. 6. A Frumkin-Fowler's adsorption isotherm for uniform surfaces. **B, C** Adsorption isotherm for heterogeneous surfaces with condensation [after Ref. 89]

sites of this fraction are taken after condensation between $\Delta\mu$ and $\Delta\mu + d\Delta\mu$ the adsorption isotherm changes from Θ to $\Theta + d\Theta$ ($d\beta = d\Theta = -\Theta' \, d\Delta\mu$). The shape of such an isotherm is shown in Fig. 6B, curve 1. For a high number of areas composed of sites of the same energy ($\omega > 4\,kT$) curve 1 becomes a straight line (curve 2). The shape of this curve is constant and independent of ω (i.e. of the alkyl chain length).

In relation to the area distribution we can obtain an isotherm with the shape shown in Fig. 6C. On the basis of previous equations and assumptions it is possible to derive the following equation for each area i:

$$kT \cdot \ln c = -\phi_{a,i} + kT \cdot \ln \frac{\Theta_i}{1-\Theta_i} + B \tag{42}$$

where $\phi_{a,i} = \phi^0_{a,i} + \Theta_i \cdot \omega$, $\Delta\mu^* = kT \cdot \ln c_1^*/c_0 = \phi^0_0 - \phi^0_{a,i}$, and Θ_i is the surface coverage of an i area, the normal energy of which is $\phi^0_{a,i}$. When the number of steps is large each point of the isotherm is practically connected with an elementary step corresponding to $\Theta_i = 0.5$ and a concentration c_i, so that

$$kT \cdot \ln c_i^* = -\frac{\omega}{2} + D \tag{43}$$

where D is a constant containing $\phi^0_{a,i}$. This equation can be used for calculating the attractive energy between CH_2 groups for the given Θ. Increasing chain length for a homologous series of surfactants has the same effect as for homogenous surfaces. The adsorption of a surfactant is carried out by condensation of one energetic area after another, and each saturated area is characterized by an undersaturation ($\Delta\mu_i^* = \phi^0_0 - \phi^0_{a,i}$).

Schechter et al.[96] developed a *thermodynamic predictive adsorption model* from the adsorption isotherm of isomerically pure alkylbenzene sulfonates on Al_2O_3 and kaolinite which describes the isotherm shape combined with its application to fit adsorption data. The use of isomerically pure substances is justified by the fact that surfactants as racemates cause maxima and minima on isotherms due to intercomponent interactions. None of the models for homogenous surfaces discussed in Chap. 3.2 is able to predict adsorption isotherms of the shape described by Schechter et al. Developing the adsorption isotherm they assume that the surface is composed of patches[97] with different adsorption energies; the energies are, however, uniform within a given patch at each site. This model includes the bilayer adsorption, lateral interactions and two-dimensional phase transition below the CMC with respect to surface heterogeneity.

The values of the chemical potential of the surfactant adsorbed at the second layer ($\Delta\bar{\mu}_2^0$) and the lateral interaction (ω) are supposed to be the same for all patches since surface heterogeneity should have little effect on second-layer adsorption of lateral interaction. The total adsorption at a given concentration is the area weighed sum of the adsorption on each patch:

$$\Theta_t = \int_0^1 \Theta_H \alpha A = \int_{\infty}^{-\infty} \frac{\Theta_H \exp[-(\Delta\bar{\mu}_1^0 - \Delta\bar{\mu}_1^M)^2/2\sigma^2] \cdot \alpha \Delta\bar{\mu}_1^0}{\sigma(2\pi)^{1/2}} \tag{44}$$

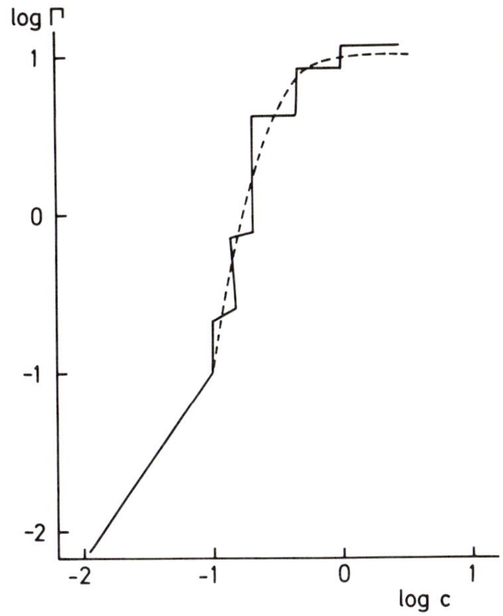

Fig. 7. Adsorption isotherm of a surfactant with patches and phase transition on heterogeneous surfaces [after Ref. 96]

where Θ_t is the total fractional coverage of the entire surface (from 0 to 2), A is the fraction of the total surface area assigned to a patch associated with the chemical potential of the adsorbed surfactant in the first layer, $\Delta\bar{\mu}_1^M$ is the mean value of $\Delta\bar{\mu}_1^0$ and σ is the standard deviation of the normalized Gaussian distribution used to describe the patch area distribution for the first layer. The resulting adsorption isotherm has a shape shown in Fig. 7 for discrete patches and a continuous distribution of patch areas. At a low surfactant concentration, $\Theta_2 \ll \Theta_1 \ll 1$ and lateral interactions are negligible on each patch; then $\Theta_t = \beta \cdot c$ and

$$\beta = K_1^A \int_{\infty}^{-\infty} \frac{\exp(-\Delta\bar{\mu}_1^0/RT)\exp[-(\Delta\bar{\mu}_1^0 - \Delta\bar{\mu}_1^M)^2/2\sigma^2]\,\alpha\,\Delta\bar{\mu}_1^0}{\sigma(2\pi)^{1/2}} \tag{45}$$

If β is constant then the Eq. 45 is transformed in the Henry's form; the adsorption is determined by the values of $\Delta\bar{\mu}_1^M$ and σ.

The course of the adsorption can be subdivided into four regions (Fig. 8) with a general configuration of the adsorbed surfactant molecules. In region I only unassociated first-layer molecules are present. The adsorption data conform with Henry's law at sufficiently low surfactant concentrations. In the second region the adsorption density increases rapidly due to lateral interactions between molecules of the adsorbed surfactant; this occurs first on the most energetic surface patches (hemi-micelles). The standard hemi-micelle is defined as a patch having an almost complete first-layer coverage ($\Theta \cong 1$) and an arbitrary second-layer coverage ($\Theta_2 < \Theta_1$) with the coverage of a given layer being identical for the different isomeric surfactant. In the regions III and IV (plateau), the surface is largely covered with hemi-micelles, with a substantial second-

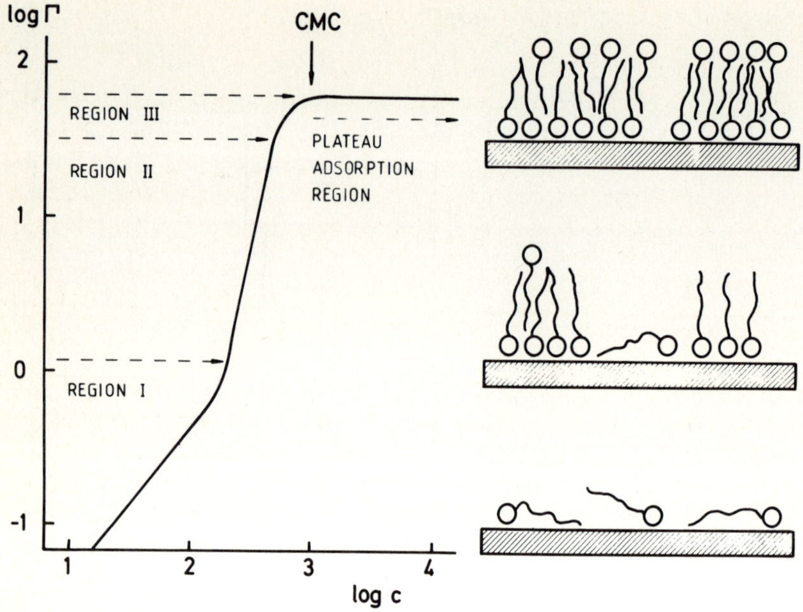

Fig. 8. Characteristic regions of adsorption with arrangement of adsorbed surfactant molecules [after Ref. 96]

layer adsorption. Above the CMC the adsorption of the surfactant is independent of the concentration. The hydrocarbon group environment in the hemi-micelle should be about the same for the first and second layers and is more favourable than that in the micelle, but a low second-layer parameter sensitivity precludes such a comparison.

Studying the adsorption of long-chain N-alkylammoniumchlorides (n_c = 8, 10, 12, and 18) on the crystalline surface of KCl, NaCl, and RbCl, or a KCl-RbCl mixture, and their saturated solutions, Matthé and Schneider[143)] obtained adsorption isotherms that can be described in a wide concentration range ($10^{-7} - 10^{-2}$ mol · l^{-1}) by Eq. 46:

$$c = \frac{k_1 \alpha}{(1 + k_2 \alpha)(1 - \alpha')} \quad \text{or in a linearized form} \quad \frac{1}{\alpha} = \frac{k_1}{c(1 - \alpha')} - k_2 \qquad (46)$$

where c is the equilibrium concentration, α and α' are the degrees of surface occupation related to a monolayer and to the apparent saturation, resp., and k_1, k_2 are constants. The constant k_1 characterizes the initial linear course of the adsorption whereas k_2 corresponds to the increment of binding character resulting from apolar interactions of surfactant chains. Plotting $1/\alpha$ against $1/c \cdot (1 - \alpha')$ we obtain a line allowing to estimate the constants k_1 and k_2 for the calculation of binding energy of apolar surfactant groups. In the range of low α values the surfactant adsorption is independent of the lattice constant and we cannot expect a selective adsorption. As the course of adsorption isotherms for mixed crystals shows, the selectivity of surfactant adsorption on alkali halides is controlled not only by hydration effects, but also by association and structural ones.

3.4 Adsorption Equation of Stern-Grahame

Several authors[98–106] describe the relation between the adsorption density of different surfactants and the electrical nature of the mineral surface using the ζ potential data.

Fuerstenau[99] was the first who used the Stern-Grahame model of EDL to describe the adsorption of long-chain surfactants for the equilibrium in heterogeneous systems. The adsorption density in the Stern plane is given by the equation

$$\Gamma_\delta = 2rC \exp\left[\frac{-\Delta G^0_{ads}}{RT}\right] \tag{47}$$

where r is the effective radius of the adsorbed ion, C is its bulk concentration and ΔG^0_{ads} is the standard free energy of adsorption $[\Delta G^0_{ads} = (\mu^0_i)^s - \mu^0_i]$. μ^0_i and $(\mu^0_i)^s$ are the standard chemical potentials of the species i in the solution and at the surface, resp.

Another approach to develop the Stern-Grahame equation is described by Fuerstenau and Raghavan[35] who apply the law of mass action to a linear mixture of similarly sized molecules showing ideal behavior in both the liquid phase and in the adsorption layer:

$$\frac{x^s_1}{1 - x^s_1} = K\frac{x_1}{x_2} = \frac{x_1}{x_2}\exp\left[\frac{-\Delta G^0_{ads}}{RT}\right] \tag{48}$$

where x_1 and x_2 are the molar fractions of the solute and solvent, resp., in the liquid phase, and x^s_1 and x^s_2 are their molar fractions in the adsorbed state. $K = x^s_1 \cdot x_2/x^s_2 \cdot x_1$ and $x^s_2 = 1 - x^s_1$. For low x_1 values the Eq. 48 is identical with the usual form of the Stern-Grahame equation and with the Langmuir equation. The equation for a surfactant adsorption in the Stern layer has a similar form in the paper of Osser-Asare et al.[100]:

$$\Gamma_\delta = \frac{\Gamma_m}{1 + 1/2\exp(\Delta G_{ads}/RT)} \tag{49}$$

where Γ_m is the adsorption density at a monolayer coverage and x is the molar fraction of the surfactant in the solution. An application of this expression involves the assumption that the size of the adsorbing species is much smaller than the radius of the adsorbent particle and that the lateral interaction between the adsorbing species is negligible. At low surface coverage this equation is simplified to the Stern-Grahame equation.

If surfactant ions adsorb only through electrostatic interactions, then $\Delta G^0_{ads} = \Delta G^0_{el} = v.F.\psi_\delta$. During specific adsorption of ions it is necessary to calculate with the additional term ΔG^0_{spec}. An estimation of this term can be made for $\psi_\delta = 0$.

For the adsorption of surfactants, ΔG^0_{ads} can be divided into a number of terms as follows:

$$\Delta G^0_{ads} = \Delta G^0_{el} + \Delta G^0_{chem} + \Delta G^0_{CH_2} + \Delta G^0_{solv} + \ldots \tag{50}$$

where ΔG^0_{el} is the electrostatic contribution to the total free energy, ΔG^0_{chem} represents the free energy due to the formation of covalent bonds with the surface, $\Delta G^0_{CH_2}$ characterizes the interaction due to the association of hydrocarbon chains of adsorbed surfactant ions, and ΔG^0_{solv} is the contribution of the solvation effects on the polar head of the adsorbate and adsorbent to adsorption. For further considerations, the mentioned authors define the specific adsorption. If the ions are adsorbed only through such forces as electrostatic attractions and through hydrophobing bonding ($\Delta G^0_{CH_2}$), the process should be termed *physiosorption.* When covalent bonds appear, the process should be termed chemisorption (the free molar energy is approximately 42 kJ or more). The adsorption density Γ_δ in the Stern plane for physisorption of surfactant ions is given by

$$\Gamma_\delta = 2r \cdot C \cdot \exp\left[\frac{-vF\psi_\delta - n\phi}{RT}\right] \tag{51}$$

where $n\phi = \Delta G^0_{CH_2}$; ϕ is the standard free energy for removing one mole of CH_2 groups from water through association and $n\phi$ is the total contribution if n is the number of CH_2 groups in the hydrocarbon chain.

Statistic mechanics were used by Kiefer and Wilson[104] to calculate adsorption isotherms of ionic surfactants on charged solid-water interface. The effect of coulombic repulsions between the ionic heads of the surfactant species are considered, as well as the van der Waals attractions of their hydrocarbon tails. Using the method of Fowler and Guggenheim[93] they obtained the equation for an adsorption isotherm:

$$\sigma = \frac{c}{c'} = \exp\left(\frac{-\chi_0}{KT}\right) \cdot \frac{\Theta}{1-\Theta} \cdot \left(\frac{2-2\Theta}{\beta+1-2\Theta}\right)^x \tag{52}$$

where

$$\beta = \{1 - 4\Theta(1-\Theta)[1 - \exp(-2w/xkT)]\}^{1/2} \tag{53}$$

and

$$c' = \left(\frac{2\pi mkT}{h^2}\right)^{3/2} \cdot kT\,\frac{j^S(T)}{j^A(T)}, \tag{54}$$

σ is the reduced bulk concentration of the surfactant, Θ is the fraction of sites with adsorbed surfactant ions, c and m are the concentration and mass of the surfactant ion, resp., h is Planck's constant, χ_0 is the increase in energy when a surfactant ion is adsorbed on an isolated site of the surface, x is the number of nearest neighbour surfactant ions in the condensed phase (sites), $j^S(T)$ is the partition function for the internal degrees of freedom of a surfactant ion in solution, and $J^A(T)$ is the same function of an adsorbed surfactant ion. The value 2 w/x corresponds to the increase in energy when a new pair of neighboring molecules of surfactant is formed and w = w (van der Waals) + w (Coulomb). The van der Waals interaction between the hydrocarbon chains is described by:

$$w(\text{van der Waals}) = -n_{CH_2} \cdot u \tag{55}$$

where $u = 4.51 \cdot 10^{-14}$ erg and n_{CH_2} is the number of CH_2 groups in the surfactant chain. The destabilizing contribution associated with the coulombic repulsion of the ionic heads is given after Stigter[120]:

$$w(\text{Coulomb}) = z_s \cdot \int_0^1 \psi_0' e \cdot d\lambda \tag{56}$$

where ψ' is the electric potential at the ionic heads of these ions when their ionic charge is $z_s \cdot e \cdot \lambda$.

3.5 Adsorption Isotherm Above the CMC

Maxima, often followed by minima, were observed on adsorption isotherms of surfactants adsorbed on solids by many authors[86, 87, 107–113]. It is not, however, possible to say that one and the same surfactant causes maxima above the CMC on all kinds of solid substances. The surface properties of minerals are one of the chief factors determining the behavior of surfactants at concentrations greater than CMC.

This was shown e.g. by investigating adsorption isotherms of Na dodecylbenzene-4-sulfonate and Na 4-hexadecyloxytolyl-2-sulfonate on various mineral surfaces differing from each other by the kind of PDI's[86]. The ζ potential value in relation to the surfactant concentration reached its maximum in the region of micelle formation and confirmed thus the shape of the adsorption isotherm. The presence of adsorption maxima is explained by a decrease in surfactant adsorption resulting from a desorption effect of micelles on the adsorption film, and by setting a three-component equilibrium (adsorption film – micelle – monomer) at concentrations $\geq$ CMC. This happens because of different ratios of the counter ions to the surfactant ions at the micelle and on the adsorption film.

Somasundaran and Hanna[129] mention two possible reasons to explain the presence of adsorption maxima: 1) exclusion of micelles form the nearsurface region of the solid due to higher electrostatic repulsion between the particles and the micelles than that between the former and singly charged surfactant monomers, and 2) alterations in solid properties such as the effective surface area, owing to change in particle morphology on excessive surfactant adsorption that can cause repulsion between various parts of the particle. Experiments of Elworthy and Mysels[192] brought evidence against the consideration of micelles as a separate phase due to decrease in the surface tension of a surfactant at concentrations above the CMC, which is not compatible with the phase separation hypothesis. To explain the formation of adsorption maxima, Sexsmith and White[112] used the law of mass action and calculated the monomer concentration in the bulk phase for the micelles with different ratios of surfactant ion number (n) to counter ion number (m). Appropriate calculations showed that variations in n and m have an influence on the concentration of monomers and that theoretical adsorption isotherms feature maxima.

The occurence of maxima on adsorption isotherms is further explained by the presence of impurities[111], being due to adsorption of laminar micelles, or by formation of hydrophobic clusters[114, 115]. The formation of maxima was criticized from the thermodynamic aspect by Kitchener[116]. Trogus et al.[117] proposed a hypothesis that minima and maxima in a total adsorption isotherm for mixtures of surfactants are predictable from considerations of monomer-micelle equilibrium relationships. According to Scamehorn et al.[96)] maxima on adsorption isotherms results from isomeric inhomogeneity of the surfactant and at a high solution-to-solid ratio.

3.6 Hemi-Micelle Formation

In the context with the course of adsorption isotherms of surfactants on solids it is necessary to mention the two-dimensional lateral aggregation between the adsorbed long-chain molecules above a given critical equilibrium concentration of the surfactant. Detailed studies of this phenomenon were published by Fuerstenau[117)], Schubert et al.[118, 119)] and Somasundaran[102, 33)]. The phenomenon is analogous to the three-dimensional micellization in the bulk phase and at the solid-water interface and is called hemi-micellization. The mechanism of hemimicelle formation is shown in Fig. 9. This two-dimensional aggregation is dependent on the chemical composition of the polar head of the surfactant, structure and length of the hydrocarbon chain, ionic strength, surface potential of the substrate, pH, and temperature. Hemi-micelles obviously contain more water in their environment than micelles since smaller fractions of the CH_2 groups are possibly removed from the water phase. The hemi-micellization at the mineral-solution

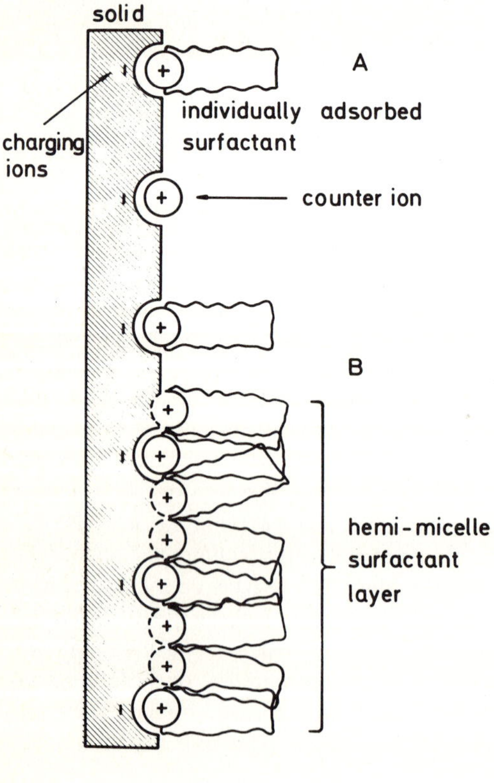

Fig. 9. Adsorbed long chain surfactant on negatively charged solid (*A*) without lateral interaction ($C \ll C_{monolayer}$) (*B*) with lateral interaction between hydrocarbon chains at higher concentrations

interface causes characteristic changes in the adsorption density, flotation recovery, ζ potential, and contact angle.

At the concentration (C_{HM}) when hemi-micelles start to form it is possible to use the Stern-Graham equation for calculation of the energy ϕ' needed to transfer a CH_2 group from the water phase to the hemi-micelle[102]:

$$C_{HM} = \frac{\Gamma'}{2r} \exp\left[\frac{W_e' - n\phi'}{kT}\right] \quad \text{or} \tag{57}$$

$$\ln C_{HM} = \ln\left(\frac{\Gamma'}{2r}\right) + \frac{W_e'}{kT} - \frac{n\phi'}{kT} \tag{58}$$

where W_e' is the electrostatic interaction and n is the number of CH_2 groups. The plot of $\log C_{HM}$ against n should be a straight line with a slope of $(-\phi/kT)$ which can be used to calculate ϕ'.

3.7 *Effect of PDI H^+ and OH^- as Dependent on pH*

Studying the effect of the concentration of PDI H^+ and OH^- on the adsorption of anionic collector at the mineral surface, Hejl[121] developed interesting relations between adsorption and pH. They are based on the Langmuir adsorption isotherm as well as on thermodynamic equilibrium[122] and the adsorption theory of Last and Cook[123]. Hejl postulates that the surface of a mineral in water is occupied with H^+ and OH^- ions reacting with the anions of a collector or their nondissociated component. If the anion of a collector is the only hydrophobizing component, then for the unoccupied mineral surface holds:

$$1 - \Theta = 1 - \Theta_{X^-} - \Theta_{OH^-} - \Theta_{H^+} \tag{59}$$

where Θ is the total occupied mineral surface and Θ_{H^+}, Θ_{OH^-}, Θ_{X^-} are parts of the mineral surface occupied with H^+, OH^- and collector anions X^-, resp. The equilibrium value of the concentration of X^- in a solution is defined by the relation

$$X^- = \frac{m \cdot K_d}{K_d + [H^+]} \tag{60}$$

where m is the analytical concentration of the collector and K_d is the dissociation constant of the acid formed by the hydrolysis of the collector. At a constant flotability of a mineral, the adsorption of the collector is described by the modified Langmuir adsorption isotherm

$$K_{X^-} = \frac{K^+}{(1 - \Theta) \cdot [X^-]} \tag{61}$$

where K^+ is the chosen degree of hydrophobicity of a mineral surface which correlates with a certain flotability of the mineral and the degree of occupation of the mineral surface by collector ions.

If a mineral surface becomes hydrophobic as a result of adsorption of a hydrolytic product of the collector, i.e. a nondissociated component, the modified Langmuir isotherm gets the form

$$K_{HX} = \frac{K^+}{(1-\Theta)\cdot[HX]} \tag{62}$$

where [HX] is the concentration of the nondissociated acid and K_{HX} is the adsorption constant. By analogy with Eq. 59 the unocuppied mineral surface is defined as

$$1-\Theta = 1-\Theta_{H^+}-\Theta_{OH^-}-\Theta_{HX} \tag{63}$$

where Θ_{HX} is the part of mineral surface occupied with the nondissociated component of the collector. The equilibrium concentration of the nondissociated acid can be calculated according to the equation

$$HX = \frac{m\cdot[H^+]}{K_d+[H^+]} \tag{64}$$

The mechanism of interaction of an anionic surfactant with a mineral surface occupied with PDI H^+ and OH^- is confirmed by the position of IP in relation to the optimum of mineral flotability. If IP of a mineral occurs at a lower pH than the flotability optimum the adsorption of the nondissociated component of the surfactant is negligible. If the condition

$$[H_i^+]^2 > 2\cdot[H_0^+]^3\cdot K_d^{-1}+[H_0^+]^2 \tag{65}$$

(where $[H_i^+]$ is the H^+ concentration in IP, $[H_0^+]$ is the H^+ concentration at which the flotability is optimal is fulfilled, the anions of the collector are considered as hydrophobizing component.

If IP of a mineral occurs at a higher pH than the flotability optimum, the nondissociated acid acts as a hydrophobizing component and it holds:

$$[H_i^+]^{-2} > [H_0^+]^{-2}+2\cdot K_d\cdot[H_0^+]^{-3} \tag{66}$$

Dixit and Biswas[124] dealt with the relationship between H^+ an OH^- concentrations and adsorption of an anionic collector on a mineral surface and its flotability. They used the system zircon-Na oleate. If Na oleate is dissolved in water the following equilibrium sets up:

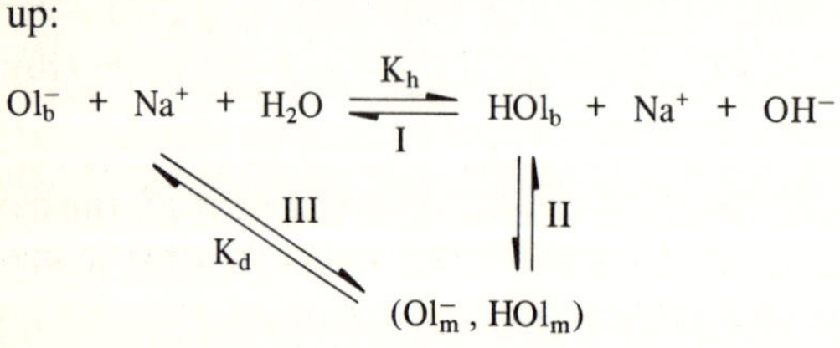

where the indices b and m belong to bulk solution and micellar phase, resp. The reactions II and III are not significant because the surfactant concentration during flotation is lower than CMC (1.5×10^{-3} mol $\cdot$ l^{-1}). The Ol^- concentration (in mol/l) in relation to pH is given by the equation

$$[Ol^-] = \frac{S[OH^-]}{K_h + [OH^-]} \tag{67}$$

where S is the actual concentration of Na oleate added in mol $\cdot$ l^{-1}. If $S \ll CMC$ the hydrolysis constant K_h is related to the degree of hydrolysis β according to the equation

$$K_h = \frac{\beta^2 S}{1-\beta} \tag{68}$$

From Eq. 67 follows that with increasing pH the Ol^- concentration increases and reaches a constant value which is dependent on the starting NaOl concentration.

For a quantitative evaluation of the effect of pH on the adsorption of a surfactant on a mineral surface the authors use the following two schemes:

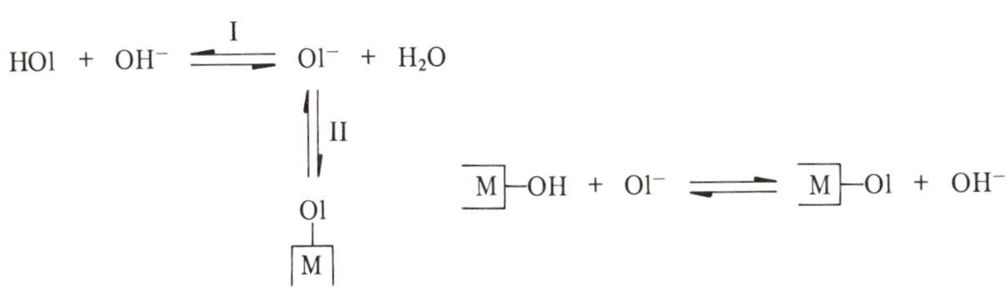

where $\boxed{M}$- represents the mineral surface. Further on they presume that the adsorption of a collector and OH^- can be described by one of the equations for the adsorption isotherm. Assuming the Freundlich adsorption isotherm, we obtain for the adsorption of Ol^- and OH^- [125])

$$\Gamma = [Ol^-]_s = k_1[Ol^-]^{n_1} - k_2[OH^-]^{n_2} \tag{69}$$

where Γ or $[Ol^-]_s$ represents the adsorption density of the collector and k_1, n_1, k_2, n_2 are empirical constants. As Fig. 10 shows Eq. 69 agrees well with experimental findings. Later studies[126)] in the same system showed that the adsorption isotherm is S-shaped and can be expressed, with the exception of very low Na oleate concentrations, by the Langmuir equation

$$\Gamma = [Ol^-]_s = \left\{\frac{S K_h K_w}{K_3(K_h[H^+] + K_w) + S K_h K_w K_4}\right\} - \left\{\Gamma_m - \left(\frac{S[H^+]}{K_c + f[H^+]}\right)\right\} \tag{70}$$

where K_w is the ion product of water and K_3, K_4, K_c, f, and Γ_m are constants which can be determined in specific cases. Dixit and Biswas[124)] also developed a relation between the flotation activity a_f and pH based on a semiempirical model of Cratin[127)]:

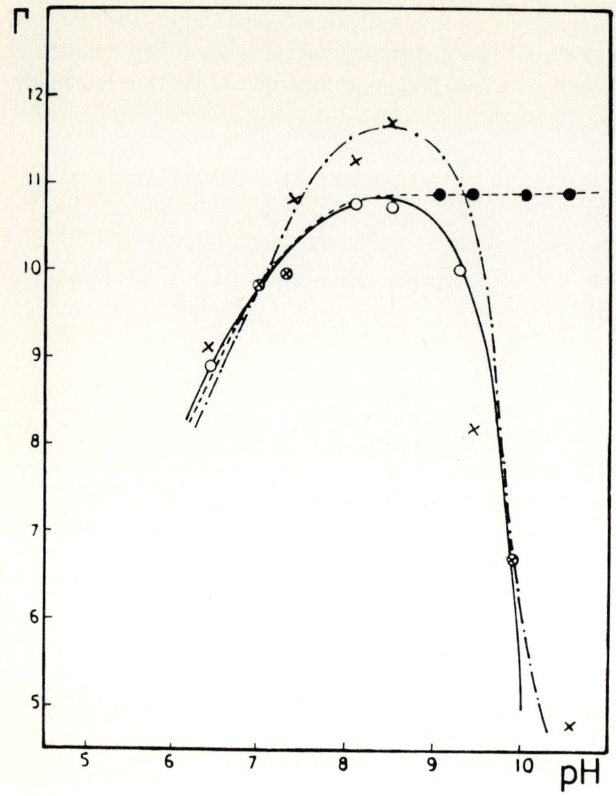

Fig. 10. Change of adsorption density Γ of Na-oleate on zircon with pH. x – experimental curve; ● – using Eq. 67; ○ – using Eq. 69 [after Ref. 124]. Γ – mol · g^{-1} · 10^7

$$R_i/R_0 = a_f =$$

$$1 - \left\{ d^c \left(\frac{1}{1 + \frac{K_w}{K_b[H^+]} + \frac{K_a K_w}{K_b[H^+]^2}} \right) + d^k \left(\frac{1}{1 + \frac{[H^+]}{K_a} + \frac{K_b[H^+]^2}{K_a + K_w}} \right) \right\} \qquad (71)$$

The flotation acitvity a_f is defined as a ratio of recovery at a given pH (R_1) to maximum recovery (R_0) ($a_f = R_i/R_0$). The constants d^c and d^k are coefficients representing the capacity of the HOl and Ol^- (bulk), resp., to affect a_f. They can be determined experimentally from the a_f vs. pH curves for different NaOl curves. The remaining constants K_a and K_b can be determined from inflex points of the a_f-pH relationship.

Akhtar and Lai[128)] infer the adsorption mechanism of a collector on a mineral surface from the mutual position of IP_{pH} and PZC which they determine by electrophoresis. Their deduction is based on experiments performed with hematite in solutions of Na oleate (NaOl) and dodecylamine hydrochloride (DDA-HCl). The surface of hematite is assumed to consist of $MeOH_2^+$, MeOH and MeO^-. Chemisorption of the collector is discussed according to Table 3 which is based on Eq. 72 for charge density at the inner Helmholtz plane:

$$\sigma_{i.H.p.} = (z_i + z_j) \cdot e\,[Me^{z_i} \cdot A^{z_j}] \tag{72}$$

where $[Me^{z_i}]$ is the concentration of surface species of the kind i in moles per unit area and $[A^{z_j}]$ is the concentration of chemisorbed species of the kind j. $z_i \cdot e$ and $z_j \cdot e$ are charges of the particular species. The difference between the PZC and IP of the solid in the presence of A^z is indicative of the chemisorption and also of the sign of $\sigma_{i.H.p.}$. Thus the sign of $\sigma_{i.H.p.}$ is determined by $(z_i + z_j)$. When z_j is known, the values of z_i can be guessed which give the correct experimentally determined sign of $\sigma_{i.H.p.}$ and thus show on what sites the A^{z_j} has chemisorbed. If z_j is assumed to be +1, 0, −1, the possible combinations of A^{z_j} with Me^{z_i} are given in Table 3, along with the expected relation of IP to PZC. Provided the assumptions made in deriving Eq. 72 are valid then, using the observed IP and ZPC from Table 3, we can judge on possible surface sites on which A^{z_j} has chemisorbed.

The variations in pH_{IP} of hematite in DDA-HCl solutions are attributed to the chemisorption of RNH_2 on MeO^- sites, and the logarithm of the equilibrium constant of the chemisorption reaction is 12.71 ± 0.14. The proposed method qualitatively accounts for the variation of pH_{IP} of hematite in NaOl. The change in the pH_{IP} occurs because of the chemisorption of RCOOH on MeO^- sites and $RCOO^-$ on MeOH sites. The chemisorption of $RCOO^-$ on MOH_2^+ sites probably occurs as well, but it does not affect the pH_{IP}.

Using the *mass-action law*, Lai and Fuerstenau[130] derived an equation to describe the distribution of charged and neutral surface sites in relation to the H^+ concentration in a solution and to the PZC of oxides. On the basis of flotation experiments they characterized the adsorption of anion- and cation-active surfactants supposing an exact correlation between the adsorption and flotability. For this purpose Na-dodecylsulfonate and dodecyl ammonium chloride were used to examine the charged sites of rutile, and Na oleate for examining the neutral surface sites. Using the charging reactions Eqs. 20 and 21[44] and the appropriate equilibrium constants $K_1 = a_{MOH} \cdot a_{H^+}/a_{MOH_2}$ and $K_2 = a_{MO^-} \cdot a_{H^+}/a_{MOH}$ they develop equations for individual fractions Θ of the surface sites of the solids:

$$\Theta_0 = \frac{K_1 (a_{H^+})}{(a_{H^+})^2 + K_1 (a_{H^+}) + K_1 K_2} \quad \text{neutral sites MOH} \tag{73}$$

Table 3. pH change of IP due to chemisorption after[128]

Surface site	Charge on chemisorbing species z_j	Expected pH of IP
$MeOH_2^+$	+1	IP > PZC
	0	IP > PZC
	−1	IP = PZC
MeOH	+1	IP > PZC
	0	IP = PZC
	−1	IP < PZC
MeO^-	+1	IP = PZC
	0	IP < PZC
	−1	IP < PZC

$$\Theta_+ = \frac{(a_{H^+})^2}{(a_{H^+})^2 + K_1(a_{H^+}) + K_1K_2} \quad \text{positive sites } MOH_2^+ \tag{74}$$

$$\Theta_- = \frac{K_1K_2}{(a_{H^+}) + K_1(a_{H^+}) + K_1K_2} \quad \text{negative sites } MO^- \tag{75}$$

The concentration of the surfactants used was below the concentration of hemimicelle formation so that they interacted in the Stern layer only with the appropriate surface site, since a physically adsorbing surfactant ion should adsorb over an opposite charged surface site. A study of the effect of pH on the relationship between flotability and the surface charge of rutile showed that the adsorption of a surfactant on a solid is well correlated with the distribution of the positive, negative, and neutral sites at the interface. The character of ion surfactant adsorption is of a physical nature whereas Na oleate chemisorbs with an optimum in the region of the PZC. An interpretation of Iwasaki et al.[131)] results of flotation measurements on goethite with a model of charged surface sites confirms a general validity of the proposed equations. The flotation limit for Na oleate related to the calculated neutral MOH site distribution agrees well with the Wadsworth's postulate that surface hydroxyls are responsible for the chemisorption of carboxylate surfactants on solids[132)].

3.8 Adsorption Kinetics

The study of adsorption kinetics of a surfactant on the mineral surface can help to clarify the *adsorption mechanism* in a number of cases. In the literature we found few communications of this kind though the adsorption kinetics has an important role in flotation. Somasundaran et al.[133, 134)] found that the adsorption of Na dodecylsulfonate on alumina and of K oleate on hematite at pH 8.0 is relatively fast (the adsorption equilibrium is reached within a few minutes) as expected for physical adsorption of minerals with PDI H^+ and OH^-. However, the system K oleate-hematite exhibits a markedly different type of kinetics at pH 4.8 where the equilibrium is not reached even after several hours of adsorption. Similarly, the effect of temperature on adsorption density varies. The adsorption density of K oleate at pH 8 and 25 °C is greater than at 75 °C whereas the opposite is true at pH 4.8. Evidently the adsorption of oleic acid on hematite involves a mechanism that is different from that of oleate or acid soaps.

Recently, Dobiáš[87)] studied the adsorption kinetics of anion- and cation-active surfactants on mineral surface with PDI H^+ and OH^- (Al_2O_3, SiO_2, TiO_2). From a previous concept, of adsorption[84)] it can be presumed that the binding of a surfactant ion on solids occurs, under simplified conditions, mostly as substitutions for a like-charged H^+ or OH^- in EDL via ion exchange or directly with charged crystalline lattice elements by ion pairing (Fig. 11). The complete adsorption layer has an apolar character.

The adsorption kinetics was examined according to the change of adsorption density in the time for the equilibrium concentration $<$ CMC. The validity of the proposed adsorption mechanism was verified by using the general kinetic equation:

$$\frac{d\Theta}{dt} = k_t(1 - \Theta) \cdot c - k_- \cdot \Theta \cdot c_{OH} \tag{76}$$

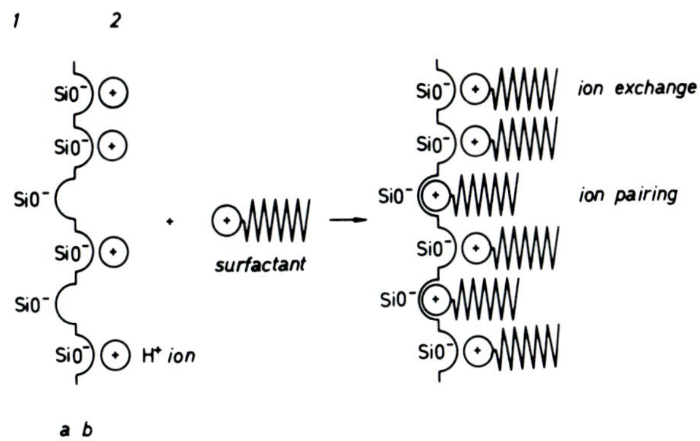

Fig. 11. Scheme of a two-step adsorption of a surfactant on a solid. *1*) solid, *2*) layer of charging ions; *a*) free surface charge, *b*) surface charge with a counterion

where Θ is the solid surface occupancy by surfactant molecules, c is the concentration of the surfactant, and c_{OH^-} (or c_{H^+}) is the OH^- (or H^+) concentration, resp., in the time t; k_+ and k_- are the rate constants of adsorption and desorption. By using this equation, two relations were derived[84] from the Langmuir and the Elovič-Zeldovič kinetic equations[135]; the latter is analogous to the Těmkin's adsorption isotherm concerning the adsorption mechanism:

$$\ln\left(1 - \frac{\Theta}{\Theta_{eq}}\right) = \frac{c_0 \cdot k_+ \cdot t}{\Theta_{eq}} \qquad \text{Langmuir equation} \qquad (77)$$

and

$$\Theta = \underbrace{\frac{RT}{b} \cdot \ln \frac{\alpha \cdot b \cdot c}{RT}}_{A} + \underbrace{\frac{RT}{b}}_{B} \cdot \ln t \qquad \text{Elovič-Zeldovič equation} \qquad (78)$$

$$\Theta = A + B \cdot \ln t \qquad (79)$$

If the solid surface is inhomogenous, k_+ and k_- are common functions of Θ and $k_+ = k_+^0 \exp(-E_+/RT)$, where E_+ is the activation energy of adsorption and k_+^0 is independent of T. Supposing E_+ is a linear function of Θ, similar to the Těmkins's equation, then $E_+ = a + b \cdot \Theta$ and a and b are independent of Θ. The significance of α is obvious considering the kinetic equation[84]:

$$\frac{d\Theta}{dt} = \underbrace{k_+^0 \cdot \exp\left(-\frac{a}{RT}\right)}_{\alpha} \cdot c \cdot (1 - \Theta) \cdot \exp\left(-\frac{b}{RT} \cdot \Theta\right) \qquad (80)$$

The adsorption kinetics according to Eqs. 77 and 78 can be verified by plotting the calculated values of $\ln(1 - \Gamma_t/\Gamma_{eq})$ against t or Γ_t against ln t. From the analysis of kinetic data it can be concluded that the Těmkin's adsorption isotherm as expressed by the Elovič-Zelovič equation, fits best to the presumed adsorption mechanism. Since the Těmkin's adsorption isotherm characterizes chemisorption, it is possible to conclude that energies derived from Coulomb's interaction during adsorption of surfactants on a mineral surface with H^+ or OH^- as PDI are comparable with energies of the chemical binding (see 3.1).

4 Adsorption Energy

4.1 Character of Binding Forces

During adsorption on a solid, surfactants differ from other adsorbates because of the polar-apolar structure of their molecule. Their adsorption from an aqueous medium on a surface of an distorted crystalline lattice is a result of a reaction equilibrium between binding forces of all species participating at this process. Because of the heterogeneity of such forces acting on the mineral-water interface it is necessary to consider several simultaneous, subsequent reactions.

According to Rosen[136] who classified the proposed models of adsorption of a monomeric surfactant on a solid, the adsorption proceeds either according to the *ion exchange mechanism*[84, 137] resulting from a replacement of counter-ions adsorbed onto a substrate by similarly charged surfactant ions, or by *ion pairing*[84, 137] when the surfactant ion adsorbs onto oppositely charged sites unoccupied by counter-ions, or by *hydrogen bonding*[137, 138] between the substrate and the adsorbate, or by *polarisation of π-electrons*[138] when the adsorbate contains electron-rich aromatic nuclei and the solid has strongly positive sites, or by *dispersion forces* of London-van der Waals[139] occuring between the substrate and adsorbate species, or finally by *hydrophobic bonding*[140, 141] resulting from an attraction between hydrophobic groups of the surfactant molecules.

In their classification Somasundaran and Hanna[129] consider the nature of the interaction forces that influence the adsorption of a surfactant. They distinguish between electrostatic (ΔG^0_{el}) and chemical (ΔG^0_{ch}) forces (ionic exchange and ion pairing according to Rosen) and newly introduce structural compatibility, hydration, and solubility factor. The standard free energy of adsorption ΔG^0_{ads} is then the sum of the contributing forces:

$$\Delta G^0_{ads} = \Delta G^0_{elec} + \Delta G^0_{ch} + \Delta G^0_{c-c} + \Delta G^0_{c-s} + \Delta G^0_{h} + \Delta G^0_{solv} \tag{81}$$

where (ΔG^0_{c-c}) is chain-chain interaction which includes the two-dimensional lateral aggregation between the adsorbed long-chain surfactant molecules above a given adsorption density. ΔG^0_{c-s} is the free energy due to a non-polar interaction between chains and hydrophobic sites on the solid, ΔG^0_{h} is the free energy of solvation or desolvation of species due to the adsorption process. The validity of the individual terms depends on the given system. For non-metallic interactions, electrostatic and lateral interaction forces are considered to be major factors determining the adsorption of surfactants, for sparingly soluble salt-type minerals, such as calcite, fluorite, and sulfides, the chemical term is significant.

Binding forces in the process of surfactant adsorption on the surface of ionic crystals were studied in detail by Richter and Schneider[74] who analyzed the adsorption conditions for low Θ values and values of Θ in the range of a monomolecular layer. For low Θ values the following energetic contributions may play a role:

E_1 – energy needed for dehydration of polar groups of the surfactant;
E_2 – energy needed for dehydration of the appropriate ions of the crystal surface;
E_3 – energy obtained by association of water dipols derived from dehydration according to E_1 and E_2;
E_4 – energy obtained by binding polar groups of the surfactant with ions of the heteropolar crystal surface;
E_5 – energy needed for repulsion of polar groups of the surfactant having the same charge sign;
E_6 – energy obtained at the parallel (or close to parallel) arrangement of apolar groups of surfactant molecules on the crystal surface due to their interaction with ions of the crystal lattice. It is supposed that at the perpendicular arrangement of isolated surfactant molecules, its apolar groups remain hydrated;
E_7 – energy needed for a partial dehydration of apolar chains of the surfactant molecule according to E_6;
E_8 – energy obtained from the association of water dipols after dehydration of apolar chains according to E_6.

E_4 is the process determining the contribution of adsorption energies. In principle, it is characterized by an ionic binding or by the Coulomb interaction and corresponds to the mechanism of ionic exchange or ion pairing. Experiments of e.g. Matthé and Schubert[106, 142, 143] showed that the surfactant adsorption proceeds far under its saturation concentration and can be well described by the Langmuir adsorption isotherm. This adsorption starts on active centers the number of which is very low; a substantial part of the adsorbed molecules is bound by steps, edges, and corners of the crystal, so that surface faults play an extraordinary role in the initial phases of adsorption[144]. The adsorption on active centers is connected with a considerable energetic gain; the surfactant molecules are isolated form each other and statistically distributed on the surface. In this case the term E_5 can be neglected. Fig. 12 A shows the interaction between a polar surfactant and a crystal surface and between the apolar part of the molecule and adsorbent surface ($\Theta \ll 1$).

$$E_p = -E_1 - E_2 + E_3 + E_4 + E_6 - E_7 + E_8 \tag{82}$$

If a surfactant binds on a solid through interlinks, such as hydrate bridges, it can be said, for the sake of simplicity, that the crystal surface as well as the polar and apolar parts of the surfactant remain hydrated. Because of the water dipoles, the binding results from weak Coulomb forces under releasing the energy E_9. This bonds are very labile and are not included in the total balance. With increasing surfactant concentration up to its saturation value a successive occupation of active centers occurs and the influence increases of their energetically advantageous regions of the lattice matrix on adsorption. The adsorption density of the surfactant reaches such an extent that at a certain concentration neighbouring adsorbed molecules interact with each other and a film-like adsorp-

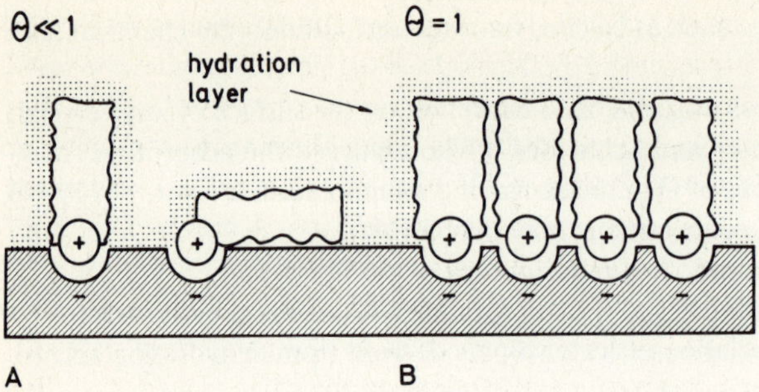

Fig. 12. Hydratation of the adsorption layer for two different surface coverage Θ [after Ref. 74]

tion film or island-like coverage appears (Fig. 12B). This arrangement of adsorbed molecules of the surfactant at the saturation threshold can be explained by the phase transition theory. This interaction process of apolar chains is connected with further changes of energy contributions:

E_{10} – energy needed for dehydration of apolar chains;

E_{11} – energy obtained during interaction of dehydrated apolar surfactant chains;

E_{12} – energy obtained through association of water dipols after dehydration of apolar groups.

The overall energetic balance, supposing the formation of an adsorption film, will be:

$$E = \underbrace{-E_1 - E_2 + E_3 + E_4 - E_5}_{E_p} \underbrace{- E_{10} + E_{11} + E_{12}}_{E_A} \quad \text{or} \tag{83}$$

$$E = E_p + E_A$$

where E_p, E_A are the binding energies of the polar and apolar interactions, resp. Under certain simplifying assumptions, discussed in detail by Richter and Schneider[74], it is possible to write Eq. 83 in the following form:

$$E = -E_1 - E_2 + E_4 + E_{12} \tag{84}$$

4.2 *Total Adsorption Energy and Bonding Energy of Hydrophobic Chains from Adsorption Isotherms*

The adsorption of surfactant on a mineral surface is generally an exothermic process and at a constant surfactant concentration in the solution its adsorbed amount decreases with increasing temperature. Using the partial derivation $\partial \ln c/\partial T$ for a constant number of CH_2 groups n of an hydrocarbon chain of a surfactant and constant values of the surface coverage α, we arrive to an equation for calculation of the total adsorption energy E. From the relation $E = E_p + E_a$, Richter and Schneider[74] derived an equation which is valid for all isotherms:

$$\left(\frac{\partial \ln c}{\partial T}\right)_{n,\alpha} = -\frac{E}{kT^2} \tag{85}$$

For α, the values of Γ or m/m_1 are introduced (see Chap. 3). The equation is verified by plotting the experimental data $\ln c$ against $1/T$; the slope of the line determines the value $-E/k$. If the experimental isotherms agree with adsorption isotherms according to Langmuir or Těmkin, it is possible to calculate the adsorption energy E using the constant β or A, resp., which were obtained from isotherms at different temperatures. E is estimated from the slope of regression lines $\ln \beta$ against $1/T$ (Langmuir) or $\ln A$ against $1/T$ (Těmkin), at a constant number of CH_2 groups of the hydrocarbon chain.

For calculations of the bonding energy of hydrophobic chains (association energy) it is possible to presume that the total adsorption energy consists of a component independent of the chain length (E_p) and components dependent on the chain length (E_A). For the term E_A, in agreement with Schubert[106, 118, 145)], it holds:

$$E_A = a \cdot n + b - k^* \cdot s \cdot \phi (n - 1) \tag{86}$$

where ϕ is the energetical gain at a complete association for a CH_2 group, s is the association degree, k^* is the pre-association coefficient ($k^* = 1$ when the ionic adsorption is presumed) and $a < 0$. With respect to this equation the authors performed a number of derivations and concluded that partial derivations of all isotherms led to the same following results:

$$\left(\frac{\partial \ln c}{\partial n}\right)_{T,\alpha} = \frac{a}{kT} = -\frac{s}{kT} \tag{87}$$

and

$$\left(\frac{\partial \ln \alpha}{\partial n}\right)_{T,c} = -\frac{a}{kT} = \frac{s}{kT} \tag{88}$$

To determine the bonding energy of hydrophobic groups, $\ln c$, or $\ln \alpha$ or $\ln \Gamma$ or m/m_1 is plotted against the number n of CH_2 groups of the homologic sequence of non-branched hydrocarbon residues of the surfactant. The slopes of the regression lines reflect the values of $-s\phi/kT$ or $s\phi/kT$. The $s\phi$ value represents the partial adsorption energy of a surfactant for an increment of one CH_2 group in the hydrocarbon chain.

The binding energy of one CH_2 group in apolar chains can also be determined by using the constants β and A from the Langmuir and Těmkin isotherm, resp. In the case of the Langmuir adsorption isotherm we obtain the following equation for two different n:

$$\ln \beta_1 - \ln \beta_2 = \ln \frac{\beta_1}{\beta_2} = \frac{a(n_1 - n_2)}{kT} \tag{89}$$

These equations are valid under the condition that A_0, β_0 and E_p from appropriate linearized equations are independent of the hydrocarbon chain length.

4.3 Determining the Binding Energy of Hydrocarbon Chains

Calculating the association energy, Schubert uses Eq. 47 (see 3.4) for the case that the adsorbed amount m (m = f(c)) of at least two homologous surfactants is known. This method allows to calculate the values of $s\phi$. For a calculation of the association degree s ($0 \to 1$), which is proportional to the arrangement of surfactant coverage on a solid, it is presumed that $\phi = -1.39\ kT$[146]. This value corresponds to a complete transfer of one CH_2 group from an aqueous medium to the environment saturated with hydrophobic chains (condensed state). Calculating the association energy, Matthé[74] uses the equation

$$\frac{c}{\alpha} = \frac{k_1}{(1 + \alpha k_2)(1 - \alpha)} \exp(E/kT) \cdot f(\alpha) \tag{90}$$

where $f(\alpha) = 1/1\text{-}\alpha$ and $\alpha = \alpha'$. On the basis of further calculations the author arrives to the equation for calculating the binding energy needed for one CH_2 group:

$$\frac{E_A}{kT} = \frac{s\phi}{kT} = -\frac{\ln(1 + k_2\alpha)}{n - 1} - \underbrace{\frac{\ln C + E_p/kT}{n - 1}}_{C'} \tag{91}$$

where C is a undefined constant, and therefore it is not possible to determine the whole term C'. The association energy can be only calculated when C' = 0.25, a value obtained according to Schubert.

To avoid the use of the auxillary value C', Matthé[74] developed another equation for calculating the association energy from the expression

$$C \cdot \frac{k_1}{1 + k_2\alpha} = \exp(E/kT) \tag{92}$$

where C is constant, so that from two different lengths of the hydrophobic chain (n_1 and n_2) it is possible to derive an equation under the assumption that the Matthé's isotherm is valid:

$$\frac{s\phi}{kT} = \ln\left(\frac{k_{11}}{k_{12}}\right) \cdot \frac{1}{n_1 - n_2} \tag{93}$$

The constant k_{11} and k_{12} correspond to the Langmuir constants for hydrocarbon chain length n_1 and n_2.

Schubert et al.[106, 118, 119], Schneider[147], and Matthé[148] analyzed the adsorption isotherms of N-alkylammonium chloride on SiO_2 in order to estimate the binding energy of hydrophobic groups. They used the constants β and A from the appropriate equations of the Langmuir and Těmkin isotherms, resp. Their results are summarized in Table 4[74]. Dobiáš[85] calculated the association energy for adsorption of $C_{10} - C_{18}$ N-alkylammonium chloride in the range of surface coverage 0.2 – 0.8 on biotite according to Schubert[119] an found values of E_A ranging from 0.401 to 0.459 kT. From these results it follows that with

Table 4. Bonding energy E_A of hydrocarbon chains for one CH_2-group

Author	equation used: $\Gamma = 2 \cdot r \cdot C \cdot \exp(-E/kT)$	from Langmuir isotherm	from Těmkin isotherm
Schneider[147] surface coverage 0.0275–0.275 (low concentration range of the surfactant)	0.44–0.64 kT n = 14 and 16	0.31 kT n = 12 and 16	0.41 kT n = 12 and 16
Schubert[106, 118, 119] surface coverage 0.2–0.8 higher concentration range of the surfactant	0.54–0.9 kT n = 12, 14 and 16	0.88 kT n = 12, 14 and 16	1.05 kT n = 12, 14 and 16
Schubert the same conditions but at a constant ionic strength (10 g KCl/l)	0.74–0.94 kT n = 12, 14 and 16	1.03 kT n = 12, 14 and 16	1.22 kT n = 12, 14 and 16
Matthé[148] low concentration range of the surfactant	equation used: $\frac{s\phi}{kT} = \ln\left(\frac{k_{11}}{k_{12}}\right) \cdot \frac{1}{n_1 - n_2}$		n = 8 and 18 → 0.23 kT n = 12 and 18 → 0.20 kT n = 8 and 12 → 0.27 kT

increasing surface coverage by surfactant molecules the association energy also increases. This binding energy value is of the order of van der Waals energy (0.008–25 kJ); whereby 1 kT = 2.5 kJ/mol. At higher concentrations of indifferent electrolytes, such as KCl of NaCl, E_A increases since the surfactant adsorption also rises as a result of binding water molecules by KCl.

4.4 Determination of Binding Energy E_p and E_A from Crystal Growth

The adsorption energy can be also determined from adsorption isotherms obtained in experiments with growing crystals in saturated solutions of water-soluble substances under a temperature gradient. The addition of a surfactant in a over-saturated solution with crystal nuclei has an influence on the growth rate of crystals resulting from its adsorption on the growth area, as a function the surfactant concentration. Richter and Schneider[74] performed such adsorption measurements at 30 °C in the system KCl-n-Alkylammonium chloride, for n = 0, 10, and 12, and arrived to a modified Langmuir adsorption isotherm, from which they derived the constants β. Association energy of the hydrocarbon chain was then calculated as discussed in 4.2 and the following values of $s\phi$ were obtained: for n = 0 and 10, −0.99 kT; for n = 10 and 12, −0.61 kT.

An interesting result is a dependence of E_A of the number of CH_2 groups in the hydrocarbon chain. This phenomenon cannot be explained by the association degree which is independent of the hydrocarbon chain length. On the basis of this finding the authors tried to calculate the energy E_p from the experimental data of Blisnakov and Nikolaev[149] who investigated the crystal growth in the system KBr – alkyl carboxylic acids (n = 0 – 4). They obtained adsorption isotherms of the Langmuir type for various hydrocarbon chain lengths on the basis of which the constants were determined and, according to 4.2, the total adsorption energy was calculated for the COOH-group. The adsorption energy for formic acid (n = 0) characterizes the polar component of the energy E_p or indicates at least its order of magnitude; it has the value of 17.33 kJ/mol = 6.86 kT. Using the Langmuir constants the binding energy of hydrocarbon chains was calculated for a CH_2 group of individual carboxylic acids ($s\phi$ = 0.22 kT). This value agrees well with the results of Matthé[148)] and Schneider[147)] (0.20–0.27 kT and 0.31 kT, resp.).

4.5 *Adsorption Heat*

From the adsorption heat of a surfactant it is possible to conclude the interaction and the mode of attachment of surfactant molecules on the crystal surface. Mellgren[150)] was one of the first investigators who analysed the adsorption heat and surface reactions of K ethyl xanthate (KEX) on galena (PbS). He showed that an ionic mechanism controls the adsorption process and that the adsorption heat of xanthate on galena is adequate to the heat reaction of xanthate with bulk Pb^{2+}. The entropy content of the "surface compound" on galena differs from that of a corresponding bulk Pb salt. The heats of KEX adsorption ($-\Delta H$) on untreated galena depend on pH values and reaction time. E.g., at pH 11.85 the $-\Delta H$ is 51.48 kJ/mol Pb^{2+} for the reaction time of 3 min, whereas for 20 min, the value is 234.41 kJ/mol Pb^{2+}. Later measurements fo Mellgren and Rao[151)] on galena and with Pb^{2+} salts using diethyl dithiocarbamate (DEDTC) were relevant to experiments using KEX. DEDTC reacts chemically with the oxidation products on galena and the anions liberated in this reaction are partly retained at the surface in the EDL, but they are rejected into the solution on excess surfactant addition. The initial heats of adsorption of DEDTC on galena are of the same order of magnitude as $-\Delta H$ associated with carbamate interaction with the appropriate Pb^{2+} salts ($-\Delta H$ = 66.55 – 111.76 kJ/mol Pb^{2+}).

The type of the oxidation product on galena is independent of the chemical environment during preparation. Rao[152)] measured the adsorption heat of K amyl xanthate (KAX) on unactivated and Cu^{2+}-activated pyrrhotite (FeS) and compared his results with heats of the reaction between KAX and Fe^{2+} or Cu^{2+} salts. With the unactivated mineral, the interaction involves a chemical reaction of xanthate with Fe^{2+} salts present at the interface (i.e. not bound to the crystal surface). The adsorption enthalpy is identical with the formation of Fe^{2+} amyl xanthate: $FeSO_4 + 2\,KAX \rightarrow FeX_2 + K_2SO_4$, and $-\Delta H$ = 97.45 kJ/mol Fe^{2+}). As revealed from the enthalpy values and the analysis of anions released into the solution, the interaction of xanthate with Cu^{2+}-activated pyrrhotite consists of xanthate adsorption by exchange for sulfate ions (formed by an oxidation of sulfides) at isolated patches (active spots), and by further multilayer formation of xanthate. The adsorption heat of KAX on pyrrhotite at the initial pH 4.5 was $-\Delta H$ (FeS unactivated) = 93.55 kJ/mol Fe^{2+} and $-\Delta H$ (FeS activated) = 70.03 kJ/mol Cu^{2+}.

Roy and Fuerstenau[153)] and Mellgren[154)] determined the adsorption heat of long-chain surfactants on oxides with PDI H^+ and OH^- and with surfactants adsorbing as counterions in the EDL. The former authors[153)] studied the immersion heat of α-Al_2O_3 in solutions of Na dodecylsulfonate (Na-DS) as dependent on pH. They found that the immersion heat is more negative when hydrocarbon chains of adsorbed Na-DS start to associate than the heat in pure water, and reaches the values of -1100 erg/cm^2 (or 1.1×10^{-4}J/cm^2) at the monolayer coverage. The value of the heat does not change even at higher coverages. On the basis of these findings Roy and Fuerstenau concluded that a layer of water molecules must be present between the adsorption layer of the surfactant and the mineral surface (term E_q at Richter and Schneider[74], see 4.1). For this model they estimated an integral heat of adsorption of -5.0 kJ/mol. Mellgren et al.[154] performed measurements on hematite in Na dodecylsulfate solutions and found a value of -5.7 kJ/mol. In contrast to Roy and Fuerstenau a constant value of the adsorption heat was reached in their experiments already at 1/6 of the monolayer coverage. To determine the total adsorption energy E and its polar component E_p, Richter et al.[155] carried out calorimetric measurements in the system N-dodecylammonium chloride (DDACl) – quartz and DDACl – KCl or NaCl. For calculating E they used the equation

$$\Delta Q_{obs} = (1 - \alpha) \cdot \Delta Q_L + \Delta Q_T \tag{94}$$

where ΔQ_{obs} is the observed heat, ΔQ_L is the heat of immersion for the solvent, ΔQ_T is the adsorption heat of the surfactant and α is the coverage of the surfactant on the solid surface. All heats are expressed as J/g. Fig. 13 shows the values of E found as dependent on α, and supplemented with values of association energy of hydrocarbon chains E_A according to Schubert et al.[119]. The curves show that the polar component of E corresponds to chemisorption and exceeds markedly the component E_A. The value of adsorption energy for $\alpha = 3\%$ in a saturated KCl solution is extremely high (879.06 kJ/mol) and is explained by a preferential adsorption of the surfactant in a solution of an indifferent electrolyte such as KCl.

Adsorption heats of DDACl on KCl and NaCl determined in saturated solutions of the electrolytes show marked differences (Fig. 14) both in their absolute values and in the signs. Whereas the surfactant (DDACl) adsorption on KCl is clearly exothermic, the reaction with NaCl is slightly endothermic. The value of E_{KCl} attains its maximum at a

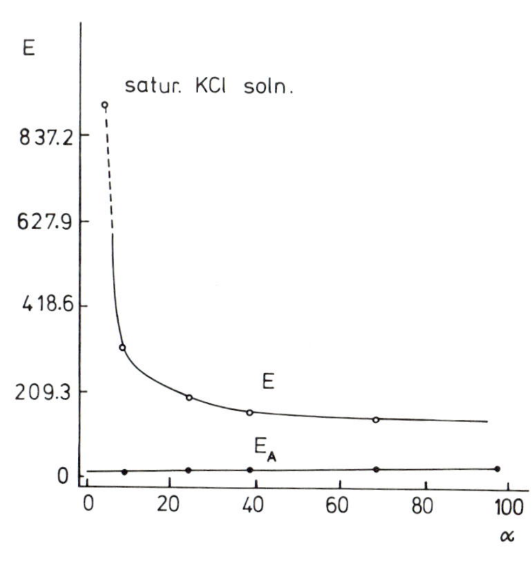

Fig. 13. Total and association adsorption energy E and E_A (in kJ/mol) of adsorbed dodedylammonium chloride on SiO_2 as dependent on surface coverage α in % [after Ref. 119 and 155]

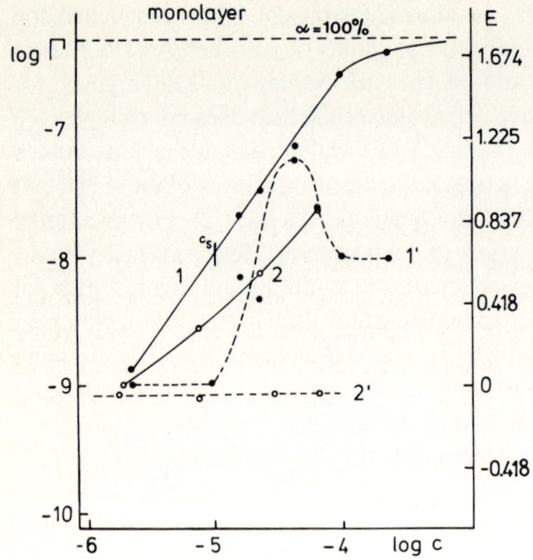

Fig. 14. Adsorption isotherms of dodecylammonium chloride (Γ in mol · g^{-1}) on KCl (1) and NaCl (2) in their saturated salt solution and appropriate heat of adsorption E (1′, 2′) in J · g^{-1} at 25 °C and pH 6 [after Ref. 155]. C_s – saturation concentration of the surfactant in saturated KCl solution

concentration, or surface coverage, that is under the value of a monolayer and confirms the dependence of E on α. The adsorption of DDACl on KCl is structure-oriented; in solutions of a higher surfactant concentration (i.e. after exceeding the saturation range in a saturated KCl solution) a heteroepitactic accumulation on the solid appears. On the contrary, no adsorption was observed of DDACl on NaCl which would be dependent on the structure.

Fuerstenau et al.[35)] mention an indirect method of determining the adsorption heat. According to this method, the adsorbed amount of a surfactant Γ is determined as a function of c (at a constant pH) at two or more different temperatures. According to the Clausius-Clapeyron type of Eq. 85 (in 4.2), the ΔH^0_{ads} is the same as the isosteric heat of adsorption Q_{st}. For surfactants such as amines and sulfonates which do not chemically react with the solid, Q_{st} is the heat of "adsorption" in the real sense of the word[35)]. In contradiction to this type on interaction they mention a chemical reaction of a surfactant such as e.g. of xanthates or salicylaldehyde[156)] when metal-surfactant complexes are formed both on the surface and in the bulk phase. In these cases Q_{st} is not the real heat of adsorption but represents an intermediate heat between the reaction heat and the adsorption heat.

The adsorption heat can be calculated using the Stern-Graham Equation, as shown by Ball and Fuerstenau[157)]:

$$\ln \Gamma_\delta = \ln 2r + \ln c - (\Delta G^0_{ads}/RT) \tag{95}$$

Differentiating this equation with respect to T and assuming r of the adsorbing ion being independent of T yields:

$$\left(\frac{\partial \ln c}{\partial T}\right) - \left(\frac{\partial \ln T}{\partial T}\right)_c = \frac{-\Delta H^0_{ads}}{RT^2} \tag{96}$$

For a constant adsorption density Γ_δ, ΔH^0_{ads} gets the isosteric adsorption heat and then it holds:

$$\left(\frac{\partial \ln c}{\partial T}\right)_{\Gamma_\delta} = \frac{Q_{st}}{RT^2} \tag{97}$$

The value of Q_{st} can be determined from the slope of the plot $(\log c)_\Gamma$ against T. Measurements in the system N-dodecylammonium acetate-quartz showed that $-\Delta G^0_{ads}$ increased with increasing temperature over the whole concentration range whereas the adsorption density against temperature showed a minimum over the greatest part of the concentration range. In the range of equilibrium surfactant concentrations of $1 - 9 \times 10^{-4}$ mol $\cdot$ l^{-1}, the values of $-\Delta G^0_{ads}$ 12.55 to 17.58 kJ/mol were found for temperatures between 5 and 45 °C.

4.6 Effect of Temperature

The course of surfactant adsorption on a mineral surface as dependent on the solution temperature is controlled, in the first place, by the character of its binding with the surface species. For the free energy change of adsorption it holds:

$$\Delta G_{ads} = \Delta H_{ads} - \Delta T \Delta S_{ads} \tag{98}$$

Unfortunately, very little is known about the entropy change (ΔS_{ads}) accompanying the adsorption of surfactants on solids. For example, Somasundaran and Fuerstenau[158)] studied the adsorption of dodecyl sulfonate on α-corundum at 25 and 45 °C and found that the adsorption density decreased with increasing temperature. A similar effect was observed by Ball and Fuerstenau[157)] who performed electrokinetic measurements of dodecylammonium acetate adsorption on SiO_2 at 5, 25, and 45 °C. In this case the absolute value of the ζ potential diminished and the PZC shifted toward higher concentrations of the surfactant with increasing temperatures, as a result of a reduction in the surface change. According to Fuerstenau and Raghavan[35)], the decrease in adsorption due to an increased temperature is typical for physisorption with an exothermic heat of adsorption. Changes in the heat and entropy of adsorption and association of long-chain surfactants at the corundum – water interface calculated from these data agree well with the conception of the two-dimensional aggregation[158)].

Quite different results were obtained by Kulkarni and Somasundaran[159)] and Raghavan and Fuerstenau[160)] who studied the adsorption of K oleate and octylhydroxamate on hematite as dependent on temperature. In both cases the adsorption of the surfactants increased with increasing temperature (25–75 °C and 20–60 °C, resp.) at low ionic strengths. At very high ionic strengths, however, the reverse was true for K oleate because of salting-out effects. By using the equation

$$\frac{\Theta}{1-\Theta} = \frac{c}{55.55} \exp\left(\frac{-\Delta G^0_{ads}}{RT}\right) \tag{99}$$

where Θ is the surface fraction covered with surfactant molecules and c is its concentration in the solution, Raghavan and Fuerstenau[160] calculated the value of –32 kJ/mol for ΔG^0_{ads}. This value is almost independent of pH and surface coverage up to $\Theta < 0.8$. The relatively low ΔG^0_{ads} value is explained with the fact that their isotherms for temperatures of 41 and 60 °C lie in the region of $\Theta > 1$. This indicated a possible formation of Fe^{3+} hydroxamate at the interface resulting from an increased solubility of the mineral at higher temperatures. The effect of the surfactant type on the course of the temperature dependence of adsorption was demonstrated by Fuerstenau and Rice[161] in experiments with pyrolusite flotation. They found that the flotation with alkylsulfonate slightly decreased with increasing temperature (23–60 °C) whereas the reverse was true with oleate in the same temperature range. At the same time the pH value diminished from 10 to 8. Dobiáš[162] studying the flotability of fluorite using Na dodecyl sulfonate, N-cetylpyridinium bromide, and ε-cyclohexenyl caproic acid found that in the temperature range from 20 to 80 °C the mineral flotability increased with rising temperature with all surfactants used. Between 50 and 60 °C, a minimum occurred which was independent of the surfactant type and which got deeper with increasing contact time of the mineral with the solution. During barite flotation with Na dodecyl sulfonate no minimum appeared. This dissimilar behavior of both minerals is explained by a better solubility of CaF_2 or its "surface compound" with increasing temperature and their different adsorption mechanisms[31]. The minimum flotability phenomenon was used for a selective separation of the mixture of CaF_2 – $BaSO_4$ – SiO_2 by using Na dodecyl sulfonate at 50 °C. Cook and Last[35], who flotated CaF_2 with oleic acid found a sudden increase in recovery and grade of fluorite at temperatures above 40 °C. They explain this observation by the change from physisorption of oleate to chemisorption at ca. 40 °C by the activation energy. From an increase in adsorption of mahogany sulfonate on sandstone with temperature Hanna and Somasundaran[163] conclude on possible chemical interactions between the sulfonate ion and the surface species such as Ca^{2+}. The original shape of the isotherm is, however, maintained in that the maximum is present at all temperatures.

5 Adsorption of Surfactants in Polycomponent Mineral Systems

5.1 Adsorption of Inorganic Ions and their Hydrolytic Products Due to Mineral Solubility

The adsorption mechanism discussed in previous chapters dealt only with a monocomponent system mineral – surfactant, as a result of an adsorption equilibrium related to appropriate PDI. Since a real flotation system consists of two or more mineral components it is necessary to mention conditions of the selective surfactant adsorption in such a system. In monocomponent systems the adsorption is controlled by the character of PDI with respect to the chemical composition of the polar heads of the surfactant. However, this rule is not valid in polycomponent systems containing both kinds of PDI. Generally and under simplifying circumstances, it is possible to classify the adsorption systems according to the role played by the PDI and the kind of the mineral.

System 1: with PDI H^+ and OH^- and the same mineral group (e.g. α-Al_2O_3–SiO_2);
System 2: with PDI H^+ and OH^- and lattice ions of different mineral groups (e.g. CaF_2–SiO_2);
System 3: with potential determining lattice ions of the same mineral group (e.g. $CaCO_3$–$Ca_{10}(PO_4)_6F_2$).

The presence of lattice ions of one component affects the solubility product of another component and this, as well as the formation of hydrolytic products of special ion species has a substantial influence on transition states between the three systems mentioned above.

System 1. Under ideal conditions the adsorption of a surfactant into the EDL proceeds as described in Chap. 3. The border of efficiency of anionic and cationic surfactants is IP or PZC, as follows from the correlation of e.g. adsorption density, potential and flotability are dependent on pH [e.g. 44, 129, 167, 174–176]. The course of such an adsorption is shown in Fig. 15. If H^+ or OH^- react with the surface of one mineral the released ions or their hydrolytic products can adsorb on unequally charged surface of the other mineral and cause an activated adsorption of the surfactant, or they can inhibit the adsorption, as shown on the schemes:

$$Fe(OH)^{2+} + [SiO_2]^- \longrightarrow [SiO_2] - Fe(OH)^+ \quad \text{activated complex} \tag{100}$$

$$[SiO_2]\text{–}Fe(OH)^+ + \ominus\!\!-\!\!- \longrightarrow [SiO_2]\text{–}\underset{\displaystyle OH}{\underset{|}{Fe^+}}\text{–}\ \ominus\!\!-\!\!- \quad \text{adsorption} \tag{101}$$

$$[SiO_2]\text{–}Fe(OH)^+ + \oplus\!\!-\!\!- \longrightarrow [SiO_2]\text{–}\underset{\displaystyle OH}{\underset{|}{Fe^+}} \rightrightarrows \oplus\!\!-\!\!- \quad \text{repulsion} \tag{102}$$

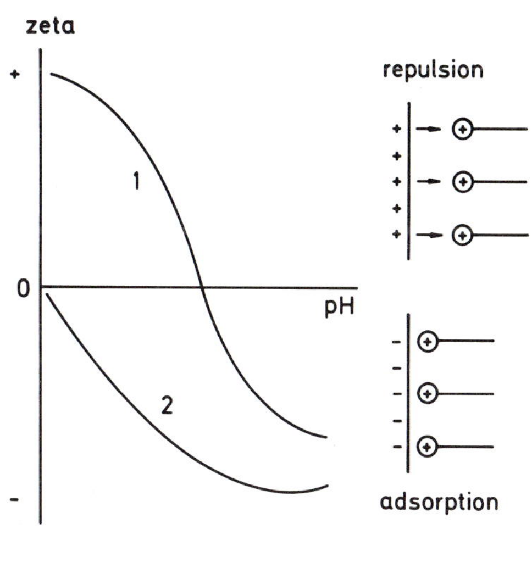

Fig. 15. Adsorption region of cationic surfactants on a mineral surface with PDI's H^+ and OH^- (1, 2) as dependent on pH and sign of the ζ potential. Using anionic surfactants, adsorption occurs on a positively charged surface and vice versa

This was shown, for example, by electrokinetic measurements of Kavanagh and Quirk[177] in the system Fe_2O_3-illite at pH 2.5. Electrophoretic and cation-exchange data indicated that the net charge of clay surface became strongly positive at low pH values as a result of adsorption of polycations Fe^{3+}. Further examples of metal ion hydrolysis in oxide and silicate flotation systems were published by Fuerstenau[178] and Stumm et al.[62].

System 2. The presence of polyvalent lattice ions in the system containing minerals with PDI H^+ and OH^- leads to their specific adsorption in the EDL and is often accompanied with a change in IP and PZC provided the surface charge has the opposite sign of that of the adsorbing lattice ion. This leads to an inhibition or activation of the mineral surface as shown in Fig. 16[179]. The same is true for a hydrolytic product of lattice ions exhibiting a stronger surface activity than non-hydrolyzed ions, as a result of a combined electrostatic and chemisorptive effect.

Studying the adsorption of lattice ions Ca^{2+} and Mg^{2+} on SiO_2, Clark and Cooke[180] reported that the adsorption isotherm of Ca^{2+} at pH 9 is similar to that of Ba^{2+}, is of the Freundlich's type, and corresponds to the formation of $Ca(OH)^+$ and $Ca(OH)_2$. The adsorption isotherm of Mg^{2+} did not show a parallelism with the isotherm of Ba^{2+} and the adsorption of Mg^{2+} appeared to correspond to the formation of very low concentrations of $Mg(OH)^+$. The ratio of Mg^{2+} and Ca^{2+} adsorption to Na^+ was calculated to be 3.91 using the Gouy-Chapman equation for the adsorption in a diffuse layer. Malati et al.[181] determined adsorption isotherms of Ca^{2+}, Sr^{2+} and Ba^{2+} on SiO_2 and found that they are of the Langmuir type and explained the adsorption mechanism in terms of electrostatic interaction between the negatively charged SiO_2 surface and hydrated cations. Dobiáš[179] who studied the adsorption of Ca^{2+} and Fe^{3+} on SiO_2 concludes that in the concentration range of 3×10^{-4} to 5×10^{-3} $mol \cdot l^{-1}$ both isotherms are of the Langmuir type whereas at lower concentrations they agree with the results of Clark and Cooke[180].

Adsorption studies with Cd^{2+}, Zn^{2+}, Cu^{2+}, and Pb^{2+} on γ-FeOOH, γ-Al_2O_3, $Fe_2O_3 \cdot H_2O$ and α-SiO_2[182] indicate that the solid surface is composed of non-uniform adsorptive binding sites. At extremely low adsorption densities, all sites are available in excess and

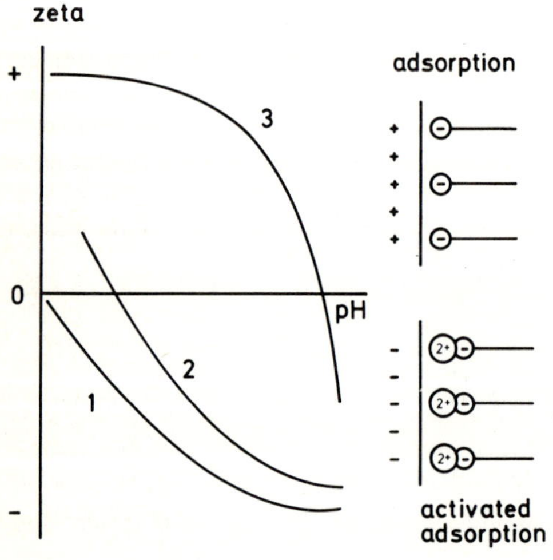

Fig. 16. Adsorption region of anionic surfactants on a mineral surface with PDI's H^+ and OH^- (*1*) and lattice ions (*3*), as dependent on pH and sign of the ζ potential; *2* – mineral 1 activated by Ca^{2+}

the surface behaves as if it were composed of identical sites. When only a few percent or less of all surface sites are occupied the strongest binding sites approach the saturation coverage and the average binding strength decreases. The average adsorption equilibrium constant decreases with increasing adsorption density. In a wide range of cation concentrations (10^{-8} – 10^{-4} mol·l^{-1}) the Freundlich's adsorption model can be used when the adsorption density is sufficiently low.

System 3. The concomitant presence of several lattice (inorganic) species and their hydrolytic products in a multicomponent system leads to a very complicated ionic composition of the EDL and reactions in the bulk phase. Moreover, at high lattice ion concentrations ion pairs and complexes can be formed. The adsorption of polyvalent ions on mineral surface with lattice PDI's was studied by a number of authors. Somasundaran[58)], Dobiáš et al.[57)], and Mishra[185)] who investigated electrokinetic properties of apatite found that adsorption of Ca^{2+} at higher concentrations leads to a sign change of the ζ potential, as a result of Ca^{2+} adsorption in the EDL. Phosphate ions in an apatite-calcite mixture adsorb preferentially on calcite and increase its negative charge. In addition, F^- were shown to alter the ζ potential of apatite significantly, probably due to precipitation of CaF_2[52)]. As reported by Bahr et al.[186)] the contact of CO_2 with the surface of fluorite leads to a formation of carbonate and thus to a change in its surface properties.

Polyvalent anions affect the adsorption of surfactants in the EDL on solids with PDI H^+ and OH^- similarly as polyvalent cations. For example, studying the adsorption of protolysable phosphate and arsenate on γ-Al_2O_3, $Al(OH)_3$, anatase and cryptocrystalline $Fe(OH)_3$ as a function of the IP_{pH}, Anderson and Malotky[183)] found that the relationship between the adsorption density Γ and the IP_{pH} can be interpreted in terms of proton dissociation occurring at, or very near to, the solid-liquid interface. On the basis of a series of Stern's equations a linearized adsorption isotherm of the Langmuir type at the IP_{pH} is used to determine Γ_{max} and free chemical energy of adsorption. The reaction of these protolysable anions with the solid surface is mainly dependent on both chemical and electrostatic forces. By interpreting the adsorption at IP_{pH}, it is possible to calculate surface dissociation constants. Experimental fit of adsorption data obtained at IP_{pH} suggests that the chemical adsorption potential does not vary with pH and implies that multiple adsorption sites or multi- vs. unidentate bonding need not be involved to model these processes quantitatively.

Studying the adsorption of NaBr on α-Al_2O_3, Smit and Holten[184)] found that the adsorption of Na^+ as well as of Br^- occurs in the compact part of the EDL. The difference between IP (pH 3.1–3.5) and PZC (pH ~ 4.5) is attributed to a greater complexation strength of Al_2O_3 for Br^- than for Na^+. The experimental data can be described in terms of site binding model of Davies et al.[70)] (see Chap. 2.3). The inner layer capacitance required to describe the pH region of predominant Br^- adsorption is approx. twice as large as that predominant for Na^+ adsorption.

5.2 *Adsorption Isotherms of Surfactants in Solutions of Multivalent Ions*

Most papers on the relation between the adsorption density of surfactant and polyvalent ions concentrate on flotation and electrokinetic evaluations[187–190)]. For example, Hopstock and Agar[187)] who studied the depressing effect of Na^+, Ca^{2+}, and Mg^{2+} on dodecylamine hydrochloride flotation of SiO_2 report that the effectiveness of these ions in inhibiting the adsorption of the surfactant is inversely related to the solubility of the hydroxide. On the basis of calculations using the Stern-Grahame equation they determined the adsorption potential ($-\phi$) of Mg^{2+} and Ca^{2+} (16.74 and 15.06 kJ/mol, resp.). Onoda and Fuerstenau[188)] investigated the effect of Na^+ and Ba^{2+} on the depression of SiO_2 flotation using dodecylammonium acetate. From electrokinetic data they gathered that the depressing effect of inorganic ions on the flotation of quartz depended on whether the surfactant ions are adsorbed as individual counterions in the EDL or whether they are associated into hemi-micelles. At low surfactant concentrations inorganic ions depress flotation through ionic competition for sites in the EDL according to the equation

$$\frac{\Gamma_a}{\Gamma_b} = \frac{c_a}{c_b} \exp\left[\frac{(G^0_{ads})_b - (G^0_{ads})_a}{RT}\right] \tag{103}$$

where Γ_a and Γ_b are the adsorption densities in the EDL of two different kinds of counter-ions and c_a, c_b are their respective bulk concentrations. If a is an ion that adsorbs only electrostatically and b adsorbs specifically, then the exponential term would simply become $\exp[(\Delta G^0_{ads})_b/RT]$. At high concentrations of the surfactant when its specific adsorption potential is high due to the association of hydrocarbon chains, inorganic ions have a small effect on flotation. Relations between the concentration of Ca^{2+} and the amount of ion adsorbed (Γ_i) on Fe_2O_3 or TiO_2 in solutions of Na dodecyltri(oxyethylene) sulfate and NaCl were studied by Murata[190)]. He found a linear increase in Γ_{surf} and $\Gamma_{Ca^{2+}}$ in relation to the $Ca^{2+}/(Ca^{2+} + Na^+)$ ratio. A similar relation for Na^+ and Cl^- yielded much smaller values of Γ. These results show that the cooperative adsorption of surfac-

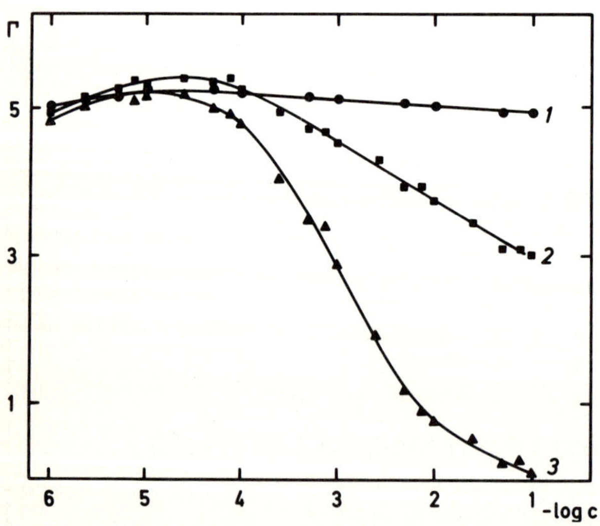

Fig. 17. Adsorption density of cetyltrimethylammonium bromide ($\Gamma \cdot 10^6$ mol $\cdot$ g^{-1}) on SiO_2 as dependent on the concentration of Na^+ (1), Ca^{2+} (2) and Fe^{3+} (3) in mol $\cdot$ l^{-1}

tants and Ca^{2+} ions on a solid occurs in the presence of other ions. Plotting the charge density (σ) values calculated from Γ_i against $[c_{Ca^{2+}}/(c_{Ca^{2+}} + c_{Na^+})]$ yields a curve with a maximum of σ due to the mutual charge cancellation of surfactant and Ca^{2+} ions.

A direct determination of the adsorption density of surfactants as dependent on Na^+, Ca^+, and Fe^{3+} concentrations was performed by Dobiáš[179]. His results are shown in Fig. 17. To interpret the results of adsorption measurement, he used the Stern-Graham equation (Eq. 51). Under the assumption that Na^+, Cl^-, Br^- and cetyltrimethylammonium ion (the surfactant used) are not the PDI, he modified the equation to get the form

$$\frac{\Gamma_{St}}{\Gamma_{Me}} = \frac{r_{St} \cdot c_{St}}{r_{Me} \cdot c_{Me}} \exp\left[\frac{F \cdot \zeta - \phi_{St} + \phi_{Me}}{RT}\right] \tag{104}$$

where Γ_{St} and Γ_{Me} are the adsorption densities of the surfactant and the polyvalent ion, resp., and ϕ_{St} and ϕ_{Me} are their specific adsorption potentials. For verification of the equation using experimental data, the following simplifications were made: r_{St}/r_{Me} and F/RT are constant, $\ln c_{ST}/c_{Me}$ varies linearly with the change of $\ln \Gamma_{St}/\Gamma_{Me}$ (this was proved), ϕ_{St} can be neglected since $\Theta \ll 1$ (association of hydrocarbon chains cannot be considered), and ϕ_{Me} remains unchanged within the range of concentrations of $CaCl_2$ and $FeCl_3$ during the adsorption, as confirmed by the linearized form of the adsorption isotherms of Ca^{2+} and Fe^{3+}. Because of a low coverage of SiO_2 surface by Ca^{2+} and Fe^{3+} ions, a homogeneous surface of the solid is assumed. A linear relationship between $\ln \Gamma_{St}/\Gamma_{Me}$ and ζ ($\zeta = \psi_\delta$) confirmed the exchange mechanism of the adsorption in the EDL without any lateral interaction for $\Theta \ll 1$.

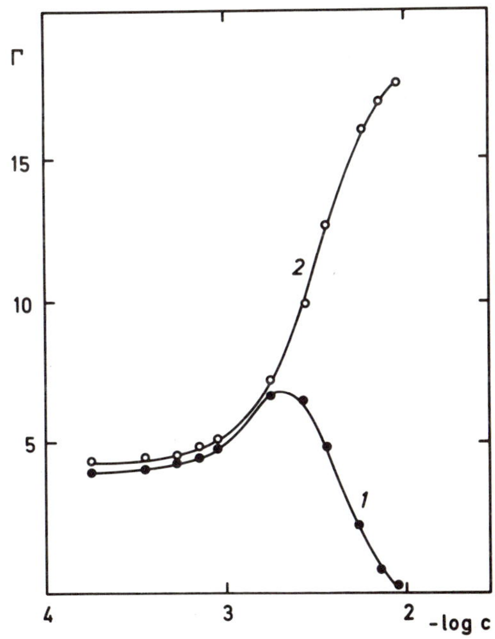

Fig. 18. Adsorption density of Na-tridecylate (NaTD) ($\Gamma \cdot 10^8$ mol $\cdot$ g^{-1}) on SiO_2 as dependent on the concentration of Ca^{2+} (mol $\cdot$ l^{-1}) 1 – system NaTD-Ca^{2+}-SiO_2; 2 – system NaTD-Ca^2-activated SiO_2

Measurements using Na tridecylate showed that the validity of Eq. 104 for SiO_2 activation by polyvalent cations is limited by the concentration at which insoluble Ca salts of the surfactant start to occur in the system (Fig. 18, curve 1). When the SiO_2 surface gets activated in a solution containing Ca^{2+} at first, and the solution is then replaced by a surfactant solution, Ca salts do not precipitate and the adsorption isotherm has its expected form (Fig. 18, curve 2). The activation of quartz by Ca^{2+}, Sr^{2+}, and Ba^{2+} ions in oleate solutions was examined by Estefan and Malati[191]. As follows from the temperature dependence (at 10, 30, and 45 °C) at pH 10.5 on the SiO_2 surface in the absence and presence of activated Ca^{2+} ions, the adsorption of Na oleate is an exothermic process and is of a physical nature. Assuming the low surface coverage, the heat of adsorption from the Langmuir adsorption isotherm was found to be 9 kJ for the activation reaction. At 30 °C the increase in the oleate adsorption density at a constant Ca^{2+} concentration of 2×10^{-4} mol/l with increasing Na oleate concentration ($1 - 4 \times 10^{-5}$ mol/l) suggests lateral interactions between the adsorbed molecules. The adsorption density of oleate increases in the order $Ca^{2+} < Sr^{2+} < Ba^{2+}$.

6 Abbreviations Used in this Paper

CFSE	Crystal Field Stabilization Energy
CMC	Critical micelle concentration
DDACl	N-Dodecylammonium Chloride
DEDTC	Diethyl Dithiocarbamate
EDL	Electric Double Layer
GChSG	Gouy/Chapman/Stern/Grahame
IP	Isoelectric Point
KEX	K Ethylxanthate
KAX	K Amylxanthate
Na-DS	Na Dodecylsulfonate
PDI	Potential-Determining Ions
PDL	Porous Double Layer
PZC	Point of Zero Charge

7 References

1. Dobiáš, B., Spurný, J.: Theory of Flotation, Chem. listy *53*, 209 (1959)
2. Schubert, H.: Aufbereitung fester mineralischer Rohstoffe Band. II, Leipzig: VEB Deutscher Verlag für Grundstoffindustie, 1977
3. Schulze, H. J.: Physikalischchemische Elementarvorgänge des Flotationsprozesses, Berlin: VEB Deutscher Verlag der Wissenschaften, 1981
4. Leja, J.: Surface Chemistry of Froth Flotation, New York: Plenum Press, 1982
5. Schneider, F.-U., Hoberg, H.: 100 Jahre Erstes Flotationspatent, Bad Harzburg, GDMB, (1977) p. 95
6. Dobiáš, B.: Tenside-Detergents 12, 210 (1975); Rudy (Prague) *12*, 397 (1964)
7. Kittabe, S.: J. Coll. Interface Sci. *48*, 327 (1974)
8. Kulkarni, R. D., Somasundaran, P.: Int. J. Mineral Process. *4*, 89 (1977)
9. Smith, R. W., Trivedi, N.: Transaction AIME *255*, 69 (1974)
10. Dorst-Hansen, W.: Ind. Eng. Chem. *61*, 10 (1969)
11. Schufle, J. A., Huang, Ch. T., Dorst-Hansen, W.: J. Coll. Interface Sci. *54*, 184 (1976)

12. Delahay, P.: Double Layer and Electrode Kinetics, New York: John Wiley (1966) p. 33
13. Shaw, D. J.: Introduction to Colloid and Surface Chemistry, London: Butterworths, (1980) p. 148
14. Stern, O.: Z. Electrochem. *30*, 508 (1924)
15. Grahame, D. C.: Chem. Rev. *41*, 441 (1947)
16. Dobiáš, B.: Erzmetall *26*, 173 (1973)
17. Hunter, R. J., Alexander, A. E.: J. Colloid Sci. *18*, 820 (1963)
18. Furlong, D. N., Yates, D. E., Healy, T. W.: Electrodes of conductive metallic oxides, Part B, Amsterdam: Elsevier (1981) p. 367
19. Mackenzie, J. M. W.: Minerals Sci. and Engineering *1971*, 25
20. Tadros, Th. F., Lyklema, J.: J. Electroanal Chem. *22*, 1 (1969); ibid *17*, 267 (1968)
21. Lyklema, J.: Electroanal. Chemistry and Interfacial Electrochem. *18*, 341 (1968)
22. Bérubé, Y. B., de Bruyn, P. L.: J. Colloid Interface Sci. *28*, 92 (1968)
23. Lyklema, J.: Croatica Chem. Acta *43*, 249 (1971)
24. Perram, J. W., Hunter, R. J., Wright, H. J. L.: Chem. Phys. Lett. *23*, 265 (1973)
25. Perram, J. W.: J. Chem. Soc. Faraday II *69*, 993 (1973)
26. Brett, A. C., Levine, S., Smith, A. L.: Natl. Colloid Symp. Texas 1974
27. Abendroth, R. P.: J. Phys. Chem. *76*, 267 (1972)
28. Van Lier, J. A., de Bruyn, P. L., Overbeek, J. Th. G.: J. Phys. Chem. *64*, 1675 (1960)
29. Kitchener, J. A.: Disc. Faraday Soc. *52*, 379 (1971)
30. Breeuwsma, A., Lyklema, J.: Disc. Faraday Soc. *52*, 324 (1971)
31. Dobiáš, B.: Tenside-Detergents *13*, 131 (1976); Freiberger Forsch. H. Reihe A *335*, 7 (1965)
32. Dobiáš, B.: Über die Adsorption der Tenside an Mineralien bei der Flotation. Symposium über Adsorption aus Lösungen, Arbeitstagung der Kolloidgesellschaft in Bochum, 29. Sept. bis 1. Okt. 1980
33. Somasundaran, P., Goddard, E. D.: Modern Aspects of Eletrochemistry Nr. 13, New York: Plenum Press, (1979) p. 207
34. Payman, M. A. F., Bowden, J. W., Posner, A. M.: Anst. J. Soil. Res. *17*, 191 (1979)
35. Fuerstenau, D. W., Raghavan, S.: Flotation, A. M. Gaudin Memorial Vol. M. C. Fuerstenau, Ed; New York: AIME (1976) p. 21
36. Boehm, H. P., Herrmann, M.: Z. Anorg. Allg. Chemie *352*, 156 (1967)
37. Boehm, H. P.: Kolloid-Z. und Z. Polymere *227*, 17 (1968)
38. Flaig-Baumann, R., Neuwinger, H. D., Boehm, H. P.: Fortschr. Kolloide und Polymere *55*, 7 (1971)
39. Little, L. H.: Infrared Spectra of Adsorbed Molecules, London: Academic Press (1966)
40. Boehm, H. P.: Disc. Farady Soc. *52*, 264 (1971)
41. Hair, J. L.: Infrared Spectroscopy in Surface Chemistry, New York: Dekker (1967)
42. Armistead, C. G., Tyler, A. J., Hambleton, F. H., Mitchel, S. A., Hockey, J. A.: J. Phys. Chem. *73*, 3947 (1969)
43. Jones, P., Hockey, J. A.: J. Chem. Soc. Faraday I *68*, 907 (1972); Trans. Faraday Soc. *67*, 2679 (1971)
44. Fuerstenau, D. W., Healy, T. W.: Adsorptive Bubble Separation Techniques, Chapter 6, New York: Academic Press, (1972) p. 91
45. Parks, G. A., de Bruyn, P. L.: J. Phys. Chem. *66*, 967 (1962)
46. Van Olphen, H.: An Introduction to Clay Colloid Chemistry, 2nd Ed., New York: Wiley (1977) p. 57
47. Parks, G. A.: Advances in Chemistry Series, Am. Chem. Soc. No. 67 (1967) p. 121
48. Quirk, J. P.: Nature *188*, 253 (1960)
49. Smolik, T. J.: Trans. AIME *235*, 367 (1966)
50. Deju, R. A., Bhappu, R. B.: Trans AIME, *235*, 329 (1966)
51. Deju, R. A., Bhappu, R. B.: State Bureau Mines Mineral Res., New Mexico Inst. Min. and Technol. Campus St. Socorro, New Mexico Circular 82 (1965)
52. Hanna, H. S., Somasundaran, P.: Flotation, A. M. Gaudin Memorial Vol., M. C. Fuerstenau, Ed., AIME New York, (1976) p. 197
53. Somasundaran, P., Agar, G. E.: J. Colloid Interface Sci. *24*, 433 (1967)
54. Dobiáš, B.: Colloid & Polymer Sci., in press 1983
55. Shergold, H. L.: Transaction Sect. C of Institution of Mining and Metallurgy *81*, C 148 (1972)

56. Samani, M. S., Blazy, P., Cases, J. M.: Transactions AIME *258*, 168 (1975)
57. Dobiáš, B., Spurný, J., Skřivan, P.: Rudy *8*, 407 (1960)
58. Somasundaran, P.: J. Colloid Interface Sci. *27*, 659 (1968)
59. Dobiáš, B.: Erzmetall XXI, 275 (1968)
60. Cibulka, J., Dobiáš, B.: Rudy *11*, Příloha Nr. 3, 1 (1963)
61. Somasundaran, P.: Clean Surfaces, Their Preparation and Characterization for Interfacial Studies, New York: Dekker (1972) p. 285
62. Stumm, W., Hohl, H., Dalang, F.: Croatica Chem. Acta *48*, 491 (1976)
63. Parks, G. A., de Bruyn, P. L.: J. Phys. Chem. *66*, 967 (1962)
64. Parks, G. A.: Chem. Rev. *65*, 17 (1965)
65. Parks, G. A.: Adv. Chem. Ser. *67*, 121 (1967)
66. Yoon, R. H., Salman, T., Donnay, G.: J. Coll. Interface Sci. *70*, 483 (1979)
67. Donnay, J. D. H., Ondik, H. M.: Crystal Data Determinative Tables, 3rd. ed. National Bureau of Standards, 1972
68. Wyckoff, R. W. G.: Crystal Structures, 2nd ed. New York: Interscience (1963)
69. Yates, D. E., Levine, S., Healy, T. W.: J. Chem. Soc. Faraday Trans. I *70*, 1807 (1974)
70. Davis, J. A., James, R. O., Leckie, J. O.: J. Coll. Interface Sci. *63*, 480 (1978)
71. Chan, D., Perram, J. W., White, L. R., Healy, T. W.: J. Chem. Soc. Faraday Trans. I *71*, 1046 (1975)
72. Sidorova, M. P., Dimtrieva, I. B., Golub, T. P.: Kollidnyi Zh. *41*, 488 (1977)
73. Sidorova, M. P., Ermakova, L. E., Kaigorodova, V. D., Tasev, D. K.: Kolloidnyi Zh. *41*, 495 (1977)
74. Richter, E., Schneider, H. A.: Freib. Forschungs.-H. A*564*, 31 (1977)
75. Jackson, K. A.: Proc. Fourth Int. Conf. on Crystal Growth, Tokyo, Japan, 24.–29. 3. 1974
76. Boudriot, H., Matthé, P., Schneider, H. A.: Freib. Forschungs.-H. A*546*, 25
77. Boudriot, H., Schneider, H. A., Matthé, P.: Neue Bergbautechnik *1*, 946 (1971)
78. Volmer, M.: Phys. Z. *22*, 646 (1921) and Volmer, M., Estermann, I.: Phys. Z. *7*, 13 (1921); Trans. Faraday Soc. *28*, 359 (1932)
79. Pashley, E. W.: Adv. Phys. *5*, 173 (1956)
80. Fripat, J., Chanssidon, J., Jelli, A.: Chimie-Physique des Phénomènes de Surface. Applications aux Oxydes et aux Silicates, Paris: Masson (1971) p. 16 and 80
81. Hiemenz, P. C.: Principles of Colloid and Surface Chemistry, New York: Dekker (1977) p. 306
82. Freundlich, H.: Kapillarchemie I, 4. Auflage, Leipzig: Akad. Verlagsges. Geest & Portig (1930)
83. Trapnell, B. W. M.: Chemisorption, Izdatelstvo Inostrannoj Literatury, Moskva (1958) p. 142
84. Dobiáš, B.: Freib. Forschungs.-H. A*401*, 123 (1966)
85. Dobiáš, B.: Tenside-Detergents *9*, 322 (1972)
86. Dobiáš, B.: Colloid & Polymer Sci. *255*, 682 (1977)
87. Dobiáš, B.: Colloid & Polymer Sci. *256*, 465 (1978)
88. de Boer, J. H.: The Dynamical Character of Adsorption, Oxford: Clarendon (1953)
89. Prédali, J. J., Cases, J. M.: Thermodynamics of the adsorption of collectors Proc. 10. Int. Mineral Processing Congress, Inst. Min. Met. London, Paper (1973) p. 33
90. Cases, J. M., Goujon, R., Smani, S.: Adsorption of N-Alkylamine Chlorides on Heterogenous Surfaces, AICHE Symp. Ser. 71, No. *150*, 100 (1975)
91. Cases, J. M., Mutaftschiev, B.: Surface Sci. *9*, 57 (1968)
92. Cases, J. M.: Bull. Minéral *102*, 684 (1979)
93. Fowler, R. H., Guggenheim, E. A.: Statistical Thermodynamics, London: Cambridge Univ. Press (1952) p. 429; Proc. Cambridge Philos. Soc. *32*, 144 (1936)
94. Hill, T. L.: J. chem. Phys. *14*, 441 (1946)
95. Stranski, I. N.: Z. phys. Chem. *136*, 259 (1928)
96. Scamehorn, J. F., Schechter, R. S., Wade, W. H.: J. Coll. Interface Sci. *85*, 463 (1982); ibid. 479 (1982)
97. Hsu, C. C., Wojciechowski, B. W., Rudzinski, W., Narkiewicz, J.: J. Coll. Interface Sci. *67*, 292 (1978)
98. Somasundaran, P., Healy, T. W., Fuerstenau, D. W.: J. Phys. Chem. *68*, 3562 (1964)
99. Fuerstenau, D. W.: Pure and Appl. Chem. *24*, 135 (1970)

100. Osser-Asare, K., Fuerstenau, D. W., Ottewill, R. H.: ACS Symp. Ser. No. 8 Adsorption at Interfaces, Am. Chem. Soc. (1975) p. 63
101. Sidorova, M. P., Kibirova, N. A., Dimitrieva, J. B.: Kolloidnyi. Zh. *41*, 277 (1977)
102. Somasundaran, P., Ananthapadmanabhan, K. P.: Solution Chemistry of Surfactants, Vol. 2, Mittal, K. L. ed., New York: Plenum (1979) p. 777
103. Sidorova, M. P., Fridrikhsberg, D. A.: Kolloidnyi Zh. *39*, 1108 (1976)
104. Kiefer, J. E., Wilson, D.: Separation Sci. and Technology *15*, 57 (1980)
105. Mukai, S., Kano, G., Matsubara, T.: Memoirs Fac. Eng., Kyoto Univ. Vol. XXI, Part 1, (1959) p. 40
106. Schubert, H.: Freib. Forschungs.-H. A*514* (1972)
107. Mukerjee, P., Anavil, A.: Adsorption at Interfaces, Mittal, K. L., Ed., ACS Symp. Ser. No. *8*, 107 (1975)
108. Gotshal, Y., Rebenfeld, L., White, H. J.: J. Coll. Interface Sci. *14*, 619 (1959)
109. Tamamushi, B., Tamaki, K.: Proc. Sec. Int. Congr. Surface Activity London, Vol. III, (1957) p. 449
110. Elworthy, P. H., Mysels, K. J.: J. Coll. Interface Sci. *21*, 331 (1966)
111. Hanna, H. S.: Contribution to the Flotation of Phosphate Ores. Ph. D. Thesis, Ain Shams Univ. Cairo 1968
112. Sexsmith, F. H., White, H. J.: J. Coll. Interface Sci. *14*, 630 (1959)
113. White, H. J.: in Cationic Surfactants, (E. Jungermann, ed.) New York: Dekker (1970) p. 311
114. Gaudin, A. M., Fuerstenau, D. W.: Trans AIME *202*, 958 (1955)
115. Cook, M. A.: J. Colloid Sci. *28*, 547 (1968)
116. Kitchener, J. A.: J. Photogr. Sci. *13*, 152 (1965)
117. Trogus, F. J., Schechter, R. S., Wade, W. H.: J. Coll. Interface Sci. *70*, 293 (1979)
118. Schubert, H., Schneider, W.: Proc. VIII Int. Congress Mineral Dressing, Leningrad, S-9, 1 (1968)
119. Schubert, H., Baldauf, H.: Tenside-Detergents *4*, 172 (1967)
120. Stigter, D.: J. Coll. Interface Sci. *47*, 473 (1974)
121. Hejl, V.: Rudy *13*, 294 (1965)
122. Modi, H. J., Fuerstenau, D. W.: J. Phys. Chem. *61*, 643 (1957)
123. Last, G. A., Cook, M. A.: J. Phys. Chem. *56*, 639 (1952)
124. Dixit, S. G., Biswas, A. K.: Adv. in Interfacial Phenomena of Particulate/Solution/Gas Systems; Applications to Flotation Research. AICHE Symp. Ser. 150, Vol. 71 (1975) p. 88
125. Dixit, S. G., Biswas, A. K.: Trans. AIME *244*, 173 (1969)
126. Dixit, S. G., Biswas, A. K.: Inst. Min. Met. Trans. Sect. C82, C-140 (1973)
127. Gratin, P. D.: Ind. Eng. Chem. *61*, 37 (1969)
128. Akhtar, S., Lai, E. L.: Adv. in Interfacial Phaenomena of Particulate/Solution/Gas Systems; Applications of Flotation Research. AICHE Symp. Ser. 150, Vol. 71 (1975) p. 110
129. Somasundaran, P., Hanna, H. S.: Improved Oil Recovery by Surfactant and Polymer Flooding, New York: Academic Press (1977) p. 205
130. Lai, R. W. M., Fuerstenau, D. W.: Transactions AIME *260*, 104 (1976)
131. Iwasaki, I., Cooke, S. R. B., Colombo, A. F.: Flotation Characteristics of Goethite, Rep. Investig. 5593, U.S. Bureau of Mines 1960
132. Peck, A. S., Raby, L. H., Wadsworth, M. E.: Trans. SME/AIME *253*, 301 (1966)
133. Somasundaran, P.: Ph-D. thesis, Univ. of California, Berkeley 1964
134. Kulkarni, R. D., Somasundaran, P.: AIME Annual Meeting, New York, Febr. 1975
135. Zeldovič, I.: Acta Physicochemika, USSR *1*, 961 (1934)
136. Rosen, M. J.: J. Am. Oil Chem. Soc. *52*, 431 (1975)
137. Rupprecht, H., Liebel, H.: Koll. Z. Z. Polym. *250*, 719 (1972)
138. Snyder, L. R.: J. Phys. Chem. *72*, 489 (1968)
139. Law, J. P., Kunze, G. W.: Soil Sci. Soc. Am. Proc. *30*, 321 (1966)
140. Schubert, H.: Rudarsko-Metalurški Zb. No. *3/4*, 225 (1967)
141. Dick, S. G., Fuerstenau, D. W., Healy, T. W.: J. Coll. Interface Sci. *37*, 595 (1971)
142. Matthé, P.: Zum Mechanismus der selektiven Adsorption von N-Alkylammoniumchloriden an Alkalihalogeniden aus wäßrigen Lösungen, Dissertation, Bergakademie Freiberg, 1971
143. Matthé, P., Schneider, H. A.: Freib. Forschungs.-H. A*564* (1977)
144. Hauffe, K., Morrison, S. R.: Adsorption, Berlin: de Gruyter (1974) p. 107

145. Schubert, H.: VI. Int. Kongr. für grenzflächenaktive Stoffe, Zürich, Bd. II, (1972) p. 172
146. Mukerjee, P.: J. Colloid Interface Sci. *XXX*, 241 (1967)
147. Schneider, W.: Dissertation, Bergakademie Freiberg, 1968
148. Matthé, P.: Dissertation, Bergakademie Freiberg, 1971
149. Bliznakov, G., Nikolaeva, R.: Kristall und Technik *2*, 161 (1967)
150. Mellgren, O.: Trans. AIME, *46* (1966)
151. Mellgren, O., Rao, S. R.: Trans. Sect. C., Inst. Min. Metall. Vol. 77, C65 (1968)
152. Rao, S. R.: Transaction AIME Vol. *250*, 199 (1971)
153. Roy, P., Fuerstenau, D. W.: J. Coll. Interface Sci. *26*, 102 (1968)
154. Mellgren, O., Gochin, R. J., Shergold, H. L., Kitchener, J. A.: Proc. 10th. Int. Mineral Process. Congress, London (1973) p. 541
155. Richter, E., Schneider, H. A., Wolf, G.: Freib. Forschungs.-H. A593 (1978)
156. Rinelli, G., Marabini, A. M., Alesse, V.: Flotation, A. M. Gaudin Memorial Vol., (M. C. Fuerstenau Ed.) New York: AIME (1976) p. 549
157. Ball, B., Fuerstenau, D. W.: Disc. Faraday Soc. *52*, 361 (1971)
158. Somasundaran, P., Fuerstenau, D. W.: Trans. AIME *252*, 275 (1972)
159. Kulkarin, R. D., Somasundaran, P.: 104th. AIME Annual Meet. New York, 1975. Oleate Adsorption at Haematite/Solution Interface and its role in Flotation
160. Raghavan, S., Fuerstenau, D. W.: Disc. Faraday Soc. *52*, 361 (1971)
161. Fuerstenau, M. C., Rice, D. A.: Trans. AIME *241*, 453 (1968)
162. Dobiáš, B.: Chem. listy *52*, 1 (1958) and Coll. Czechoslow. Chem. Commun. *24*, 3029 (1959)
163. Hanna, H. S., Somasundaran, P.: Improved Oil Recov. by Surfactant and Polymer Flooding, New York: Academic Press (1977) p. 253
164. Dorst-Hansen, W.: On the structure of water near solid interfaces and the possible existence of long-range-order, Investig. Rep., Inst. of Marine Sci., Univ. of Miami, 1969
165. Diemen, A. J. G., Stein, H. N.: J. Coll. Interface Sci. *67*, 213 (1978)
166. Dobiáš, B.: Erzmetall *XXI*, 275 (1968)
167. Dobiáš, B., Spurný, J., Freudlová, E.: Coll. Czechoslow. Chem. Commun. 24, 3668 (1959)
168. Vizwanathan, K. V., Shukla, S. K., Majumdar, K. K.: Trans. Indian Inst. Metals June 1970, G
169. Williams, D. J. A., Williams, K. P. J.: Coll. Interface Sci. *65*, 79 (1978)
170. Plitt, L. R., Kim, M. K.: Trans. AIME *256,* 188 (1974)
171. Atademir, M. P., Kitchener, J. A., Shergold, H. L.: J. Coll. Interface Sci. *71*, 466 (1979)
172. Tari, J., Hirai, T.: Denki Kagaku *46*, 544 (1978)
173. Healy, T. W., Moignard, M. S.: Flotation, A. M. Gaudin, Memorial Vol., (M. C. Fuerstenau, ed.) Am. Inst. Min. Metall. Petrol Eng. New York, Vol. 1, (1976) p. 275
174. Dobiáš, B., Spurný, J., Cibulka, J.: Coll. Czechoslow. Chem. Commun. *31*, 166 (1966)
175. Dobiáš, B.: Proc. IVth. Int. Congr. on Surface Active Substances, Brussels, 1964 New York: Gordon & Breach (1967) p. 681
176. Fuerstenau, D. W., Healy, T. W., Somasundaran, P.: Trans. AIME *229*, 321 (1964)
177. Kavanagh, B., Quirk, J. P.: Geoderma *21*, 225 (1978)
178. Fuerstenau, M. C.: Adv. Interfacial Phenomena of Particulate/Solution/Gas Systems; Applications to Flotation Research, AICHE Symposium Ser. 150, Vol. 71 (1975) p. 16
179. Dobiáš, B.: Berichte VI. Int. Kongr. grenzflächenaktiver Stoffe, Zürich 1972, München: Hanser (1973) p. 563
180. Clark, S. W., Cooke, S. R. B.: Trans. AIME *241*, 334 (1968)
181. Malati, M. A., Yousef, A. A., Estefan, S. F.: Chimie et Industrie-Genie Chimique *103*, 1347 (1970)
182. Benjamin, M. M., Leckie, J. O., in: Contaminants and Sediments, Vol. 2, (R. A. Backer ed.) Ann Arbor. Sci. Publ. Ann Arbor (1980) p. 305
183. Anderson, M. A., Malotky, D. T.: J. Coll. Interface Sci. *72*, 413 (1979)
184. Smit, W., Holten, C. L. M.: J. Coll. Interface *78*, 1 (1980)
185. Mishra, S. K.: Int. J. Mineral Process. *5*, 69 (1978)
186. Bahr, A., Clement, M., Surmatz, H.: 8th. Int. Mineral Process. Congress, Leningrad, Paper S-11 (1968)
187. Hopstock, D. M., Agar, G. E.: Trans. AIME *241*, 466 (1968)
188. Onoda, G. Y., Fuerstenau, D. W.: Proc. 7th. Int. Mineral Process. Cong., New York (1964) p. 301

189. Modi, H. J., Fuerstenau, D. W.: Trans. AIME *217*, 381 (1960)
190. Murata, M.: J. Chem. Soc. Japan, Chem. and Industrial Chemistry (1979) p. 984
191. Estefan, S. F., Malati, M. A.: Trans. Sect. C, Int. Min. Metall. *82*, C237 (1973)
192. Elworthy, P. H., Mysels, K. J.: J. Coll. Interface Sci. *21*, 331 (1966)

Received 8. 10. 1982
C. K. Jørgensen (editor)

Author-Index Volumes 1–56

Structure and Bonding

Editors: **M. J. Clarke, J. B. Goodenough, P. Hemmerich J. A. Ibers, C. K. Jørgensen, J. B. Neilands, D. Reinen, R. Weiss, R. J. P. Williams**

Volume 50

Topics in Inorganic and Physical Chemistry

1982. 48 figures, 29 tables. V, 178 pages. ISBN 3-540-11454-8

Contents: *J. A. Ibers, L. J. Pace, J. Martinsen, B. M. Hoffman:* Stacked Metal Complexes: Structures and Properties. – *M. J. Clarke, P. H. Fackler:* The Chemistry of Technetium: Toward Improved Diagnostic Agents. – *R. J. P. Williams:* The Chemistry of Lanthanide Ions in Solution and in Biological Systems. – *C. K. Jørgensen, R. Reisfeld:* Uranyl Photophysics.

Volume 51

A. Schweiger

Electron Nuclear Double Resonance of Transition Metal Complexes with Organic Ligands

1982. 47 figures, 19 tables. VIII, 128 pages. ISBN 3-540-11072-0

Contents: Introduction. – ENDOR-Instrumentation. – Analysis of ENDOR Spectra. – Advances ENDOR Techniques. – Interpretation of Hyperfine and Quadrupole Data. – Discussion of the Literature. – Concluding Remarks. – Appendix A: Abbreviations Used in this Paper. – Appendix B: Second Order ENDOR Frequencies. – Appendix C: Relations Between Nuclear Quadrupole Coupling Constants in Different Expressions of H_Q (Sect. 5.2). – References. – Subject Index.

Volume 52

Structures versus Special Properties

1982. 90 figures, 19 tables. V, 202 pages. ISBN 3-540-11781-4

Contents: *R. G. Woolley:* Natural Optical Activity and the Molecular Hypothesis. – *L. Banci, A. Bencini, C. Benelli, D. Gatteschi, C. Zanchini:* Spectral-Structural Correlations in High-Spin Cobalt (II) Complexes. – *A. Tressaud, J.-M. Dance:* Relationships Between Structure and Low-Dimensional Magnetism in Fluorides. – *V. K. Jain, R. Bohra, R. C. Mehrotra:* Structure and Bonding in Organic Derivatives of Antimony (V).

Springer-Verlag
Berlin
Heidelberg
New York
Tokyo